OECD Review of Fisheries: Policies and Summary Statistics 2015

This work is published under the responsibility of the Secretary-General of the OECD. The opinions expressed and arguments employed herein do not necessarily reflect the official views of OECD member countries.

This document and any map included herein are without prejudice to the status of or sovereignty over any territory, to the delimitation of international frontiers and boundaries and to the name of any territory, city or area.

Please cite this publication as:
OECD (2015), *OECD Review of Fisheries: Policies and Summary Statistics 2015*, OECD Publishing, Paris.
http://dx.doi.org/10.1787/9789264240223-en

ISBN 978-92-64-24016-2 (print)
ISBN 978-92-64-24022-3 (PDF)

Series: OECD Review of Fisheries: Policies and Summary Statistics
ISSN 2225-4315 (print)
ISSN 2225-4323 (online)

The statistical data for Israel are supplied by and under the responsibility of the relevant Israeli authorities. The use of such data by the OECD is without prejudice to the status of the Golan Heights, East Jerusalem and Israeli settlements in the West Bank under the terms of international law.

Photo credits: Cover © Shutterstock/Cienpies Design.

Corrigenda to OECD publications may be found on line at: *www.oecd.org/about/publishing/corrigenda.htm*.

Foreword

This edition of the *OECD Review of Fisheries* provides updated statistics and information on developments in policies and activities in the fishing and aquaculture sectors of OECD countries and partner economies. It covers mainly the time period 2012-13 but reports on more recent developments as well, notably changes to the Common Fisheries Policy (CFP) of the European Union, the rising importance of the concept of ocean governance, and the evolution of trade agreements and regional fisheries management organisations.

Part I provides a broad view of activities in the sector. It identifies trends in production and trade, including an outlook for future developments. Part II contains two-page country snapshots of summary statistics and key developments in the fisheries and aquaculture sectors in OECD countries, Argentina, People's Republic of China, Chinese Taipei, Indonesia, and for the first time, Latvia. The electronic version of the *Review* provides additional detail on the institutional and policy background, downloadable charts and data, and much more (available at the OECD iLibrary, www.oecd-library.org). Part II is based on contributions by participating OECD countries and partner economies.

This *Review* is authored by the staff of the OECD Fisheries Policies Division (Lae Hyung Hong, Roger Martini, Claire Delpeuch, and Carl-Christian Schmidt), based on contributions provided by participating OECD countries and partner economies.

Table of contents

Tables

Figures

Acronyms

ABNJ	Areas Beyond National Jurisdiction
ADEP	Republic of South African Aquaculture Development and Enhancement Programme
BRPs	Biological Reference Points
CDI	Center for Development and Innovation of Wageningen University
CFP	Common Fisheries Policy of the European Union
COFI	Committee of Fisheries
DAC	OECD Development Assistance Committee
DNOCS	Brazilian National Department of Works Against Drought
DTU	Danish Technology Insitutue Aqua
E-BCD	Electronic Bluefin Tuna Catch Document
EEZs	Exclusive Economic Zones
EFTA	European Free Trade Association
EMFF	European Maritime and Fisheries Fund
EPAs	Economic Parnership Agreements
EU	European Union
FAO	Food and Agriculture Organisation of the United Nations
FADs	Fish Aggregating Devices
FTA	Free Trade Agreement
GDP	Gross Domestic Product
GEF	Global Environment Facility
GFT	Government Financial Transfers
GGS	Green Growth Strategy
GPS	Global Positioning System
ICCAT	International Commission for the Conservation of Atlantic Tunas
IFSC	Brazilian Federal Institute Santa Catarina
IOTC	Indian Ocean Tuna Commission
ITQ	Individual Transferable Quotas
IUU	Illegal, Unregulated and Unreported Fishing
IVQs	Individual Vessel Quotas
MAPA	Brazilian Ministry of Agriculture, Livestock and Supply
MCS	Monitoring, Control, and Surveillance

MCTI	Brazilian Ministry of Science, Technology, and Innovation
MMO	UK's Marine Management Organization
MOMAF	Korean Ministry of Maritime Affairs and Fisheries
MPA	Brazilian Ministry of Fisheries and Aquaculture
MPEDA	Indian Marine Products Export Development Authority
MSY	Maximum Sustainable Yield
NABL	Indian Accreditation Board for Testing & Calibration Laboratories
NFDB	Indian National Fisheries Development Board
OECD	Organisation for Economic Co-operation and Development
PNCMB	Brazilian National Hygiene and Sanitary Control of Bivalve Molluscs
PO	Producer Organisations
RFMOs	Regional Fisheries Management Organisations
RGCA	Indian Central Aquaculture Pathology Laboratory of the Rajiv Gandhi Centre for Aquaculture
SIF	Brazilian Federal Inspection Services
SIOFA	Southern Indian Ocean Fisheries Agreement
SOFIA	State of World Fisheries and Aquaculture, FAO
SPRFMO	South Pacific Regional Fisheries Management Organization
SSF	Sustainable Small-scale Fisheries
TAC	Total Allowable Catches
UEMA	Brazilian State University of Maranhão
UFMG	Brazilian Federal University of Minas Gerais
UNEP	United Nations Environment Programme
VGFSP	FAO Voluntary Guidelines for Flag State Performance
VMEs	Vulneralbe Marine Ecosystems
VMS	Vessel Monitoring System
WB	World Bank

Executive Summary

There has been a continued decline in OECD fisheries production and rebuilding stocks remains a key challenge for fisheries management

Global marine capture production peaked in the 1996 and has been relatively flat or declining since that time. For OECD countries, the decline is more pronounced, with a reduction in landings of 39% since the peak in 1988. The share of OECD countries in the total world catch has decreased, from around 40% of total world catches in the late 1980s to 30% today.

Despite the OECD's moderate share of total production, global trade of fisheries products are still dominated by OECD countries. OECD countries accounted for around 72% of all imports in 2012, down from 83% in 2003. The largest consumers of fish products are the European Union, the United States and Japan.

The European Union adopted in February 2013 a new CFP which came into force on 1 January 2014. The new CFP emphasises the need to ensure sustainable fisheries and requires quotas to be set with reference to Maximum Sustainable Yield (MSY) by 2015 in most cases. The new CFP has as its objective rebuilding all fish stocks to MSY levels by 2015, or 2020 at the latest.

Following the introduction of a Total Allowable Catches (TAC) system in the early 1990s and an allocation system similar to (Individual Transferable Quotas ITQs) in 2001, Chile adopted in 2013 a new fisheries law that introduced concepts of precautionary and ecosystem approaches; new definitions and classifications for assessing and measuring the availability of fishery resources; new international sustainability management standards such as Biological Reference Points (BRPs) and Maximum Sustainable Yield.

The People's Republic of China (hereafter "China") and Indonesia alone account for nearly a quarter of global fish harvests. Asian countries dominate production, but typically, and especially in the case of China, consume the majority of production domestically. Economic growth in China has been driving an increase in consumption of fisheries products. Per capita consumption in China's cities rose from 10.34 kg in 2000 to 14.62 kg in 2011.

The centre of gravity of global fishing has moved eastward and some Asian countries are embracing integrated ocean management

China is moving toward a unified marine governance approach in order to protect their ocean interests and develop ocean-related industries. In 2013 four of its five maritime law-enforcement commands were consolidated into the SOA (State Oceanic Authority). Similarly, after a five-year period which saw the ocean portfolio split into several units, the Korean Ministry of Oceans and Fisheries was created in 2013 with responsibilities to provide a fully integrated approach to all marine issues.

India established the National Fisheries Development Board (NFDB) in August 2014 in order to have a more integrated fisheries governance system. The NFDB will promote the fisheries sector and co-ordinate activities related to fisheries undertaken by different ministries or departments in the central government and state or union territory governments.

Aquaculture now more important for human consumption than capture fisheries and gains policy attention as a result

Aquaculture has been consistently the fastest growing of all food commodities. World aquaculture production is centred in China, India, Viet Nam, Indonesia, and Bangladesh, who together account for 80% of global aquaculture production. The major OECD aquaculture producers are Norway, Chile, Japan, Korea, and the United States, who together account for 6% of global production by volume.

Indonesia in 2014 invited the WorldFish Centre to work jointly on preparing a Master Plan for Aquaculture Development until 2020. Tasks of this project include developing scenarios of supply and demand for fishery products, and building an understanding of the opportunities and challenges to stimulate sustainable aquaculture in Indonesia.

The United States set up a National Strategic Plan for Federal Aquaculture Research covering 2014 to 2019. A national strategic plan for federal aquaculture research, developed by the National Science and Technology Council's Committee on Science includes nine strategic goals with associated outcomes and milestones that identify federal agency and interagency research, science and technology priorities over five years that will support aquaculture development in the United States.

South Africa in 2013 put in place the Aquaculture Development and Enhancement Programme (ADEP) to develop the aquaculture industry by the Department of Agriculture, Fisheries and Forestry in co-operation with the Department of Trade and Industry. The goal of the ADEP is to encourage investments in the aquaculture sector to increase sustainable production and the creation of jobs and skills development.

PART I

GENERAL SURVEY OF FISHERIES POLICIES

This general survey of the Review of Fisheries describes trends in capture fisheries and aquaculture production, trade, fleet development, employment and government financial transfers, providing a snapshot of key developments. It also provides an overview of key recent policy developments. Many OECD countries and partner economies are undertaking important structural and policy reform in their fisheries sector, and new governance structures and management instruments are being put in place. Finally, this general survey offers an overview of the activities of the OECD Committee for Fisheries (COFI) and the OECD Fisheries Secretariat.

Chapter 1

Recent trends in OECD fisheries and aquaculture

This chapter describes trends in capture fisheries and aquaculture production, trade, fleet development, employment and government financial transfers, providing a snapshot of key developments. The data show the dominance of South East Asian countries in fisheries and aquaculture production and trade, and the continuation of the long-term decline in importance of OECD fisheries production. Aquaculture production will continue to be the source of growth in fish products worldwide.

The statistical data for Israel are supplied by and under the responsibility of the relevant Israeli authorities. The use of such data by the OECD is without prejudice to the status of the Golan Heights, East Jerusalem and Israeli settlements in the West Bank under the terms of international law

1.1. Marine capture fisheries

Global marine capture production peaked in the 1996 at 86.4 million tonnes and has been relatively flat or declining since that time. For OECD countries, the decline is more pronounced, with a reduction in landings of 39% since the peak in 1988. The cause of this decline is the increasing proportion of fisheries that are fully or over-exploited. Moreover, many fish stocks are in the process of recovery, which requires temporarily reducing harvest to allow the stock to recover.

On average, harvests have declined by almost 15% in the past ten years in OECD countries. However, this average hides signficant variation across countries. (Figure 1.2). The volume of catches decreased by more than half in Slovenia, Denmark, and Sweden, and most countries recorded at least some decline in harvest. Efforts to promote stock recovery are starting to bear fruit in some countries such as Korea, Ireland, the United States, Spain, Poland and Mexico. Finland recorded a 60% increase in the last decade. This is partly due to Baltic herring: the state of the Baltic herring stocks are at the moment good as quotas are set according to scientific advice and respected. Herring and sprats are two main species for Finnish marine capture fisheries. The volume of herring capture increased from a low in 2003 to its highest level ever in 2012, exceeding the previous peak in 1994.

The share of OECD countries in the total world catch has decreased, from around 40% of total world catches in the late 1980s to 30% today. The share of global harvests taken by Asian countries has increased over this period, and the centre of gravity of global fishing has moved eastward.

The seven most important producers in the OECD are the United States, Japan, Chile, Norway, Korea, Mexico and Iceland who together accounted for 75% of OECD total marine capture production in 2012 (Table 1.1). Ten of the top 18 global producers are in Asia, with the People's Republic of China (hereafter China) and Indonesia topping the list. Many of these countries have posted impressive growth in the last decade, but the question is whether increased harvest comes from greater capacity of a recovered resource or due to unsustainable fishing.

Figure 1.1. Total harvest in OECD countries has been in long-term decline

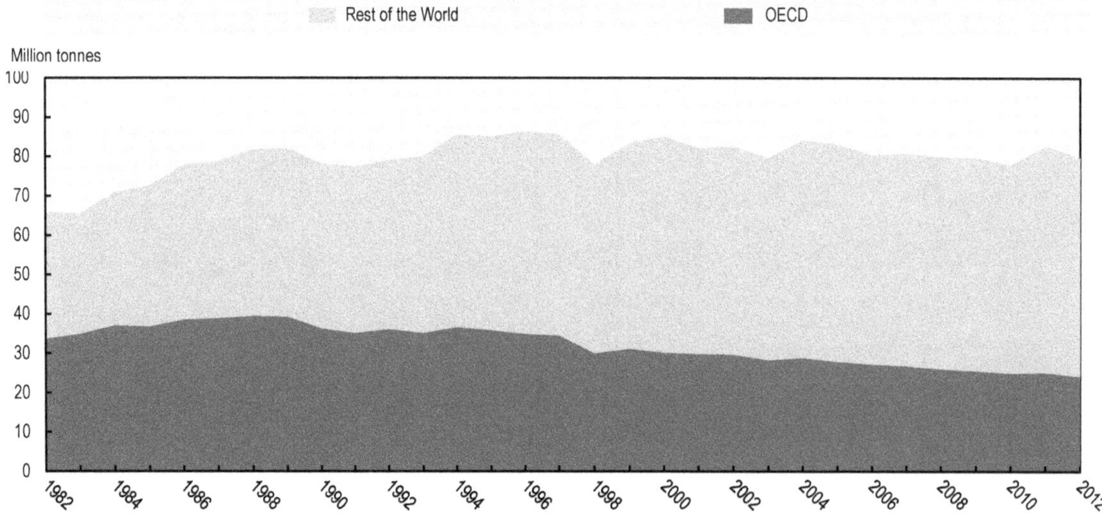

Source: OECD (2015) Fisheries and aquaculture, OECD Agriculture statistics (Database) and Food and Agriculture Organization of the United Nations. FIGIS. FishStat (Database).
dx.doi.org/10.1787/data-00285-en and www.fao.org/fishery/statistics/global-production/query/en

Variations may also be due to exogenous factors. In the case of Chile and Peru a sharp decrease of anchovy resources affected the harvest. The *El Niño* is a natural weather cycle that lowers productivity in those fisheries in certain years. Moreover, a new fisheries law in 2012 that introduced a reformed fisheries management scheme and instruments such as mandating the use of GPS mainly for large vessels (more than 12~15 m vessels), and a ban on bottom fishing as long as it is harmful for vulnerable Marine Ecosystems can be expected to impact production in the short term.

Figure 1.2. There is a broad variation in trends across OECD countries)

Percentage change in harvest (2003 to 2012

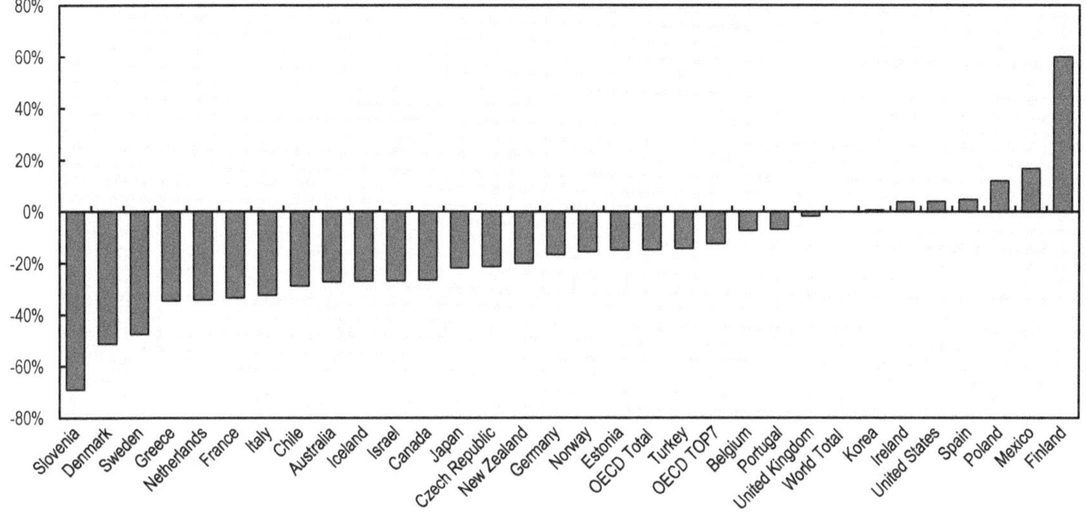

Source: OECD (2015), *Fisheries and aquaculture*, OECD Agriculture statistics and Food and Agriculture Organization of the United Nations, FIGIS. FishStat (Database), dx.doi.org/10.1787/289ec9ed-en and www.fao.org/fishery/statistics/global-production/query/en.

Table 1.1. Marine capture fisheries: Major producers (tonnes)

Ranking	Country	2003	2011	2012	'03-'12	'11-'12
1	China	12 212 188	13 536 409	13 869 604	13.6%	2.4%
2	Indonesia	4 275 115	5 332 862	5 420 247	27.0%	1.7%
3	United States	4 912 627	5 131 087	5 107 559	4.0%	-0.5%
4	Peru	6 053 120	8 211 716	4 807 923	-20.6%	-41.5%
5	Russia	3 090 798	4 005 737	4 068 850	31.6%	1.6%
6	Japan	4 626 904	3 741 222	3 611 384	-21.9%	-3.5%
7	India	2 954 796	3 250 099	3 402 405	15.1%	4.7%
8	Chile	3 612 048	3 063 467	2 572 881	-28.8%	-16.0%
9	Viet Nam	1 647 133	2 308 200	2 418 700	46.8%	4.8%
10	Myanmar	1 053 720	2 169 820	2 332 790	121.4%	7.5%
11	Norway	2 548 353	2 281 856	2 149 802	-15.6%	-5.8%
12	Philippines	2 033 325	2 171 327	2 127 046	4.6%	-2.0%
13	Korea	1 649 061	1 737 870	1 660 165	0.7%	-4.5%
14	Thailand	2 651 223	1 610 418	1 612 073	-39.2%	0.1%

Table 1.1. Marine capture fisheries: Major producers including top seven OECD countries (tonnes) (*cont.*)

15	Malaysia	1 283 256	1 373 105	1 472 239	14.7%	7.2%
16	Mexico	1 257 699	1 452 970	1 467 790	16.7%	1.0%
17	Iceland	1 986 314	1 138 274	1 449 452	-27.0%	27.3%
18	Morocco	916 988	949 881	1 158 474	26.3%	22.0%
Total 18 countries		58 764 668	63 466 320	60 709 384	3.3%	-4.3%
OECD-top 7		20 593 006	18 546 746	18 019 033	-12.5%	-2.8%
OECD-7' share of World (OECD)		25.8% (72.6%)	22.5% (73.9%)	22.6% (74.7%)		
OECD-34		28 346 747	25 098 495	24 113 070	-14.9%	-3.9%
OECD-34's share of World		35.6%	30.9%	30.3%		
World Total		79 674 875	82 609 926	79,705,910	0.0%	-3.5%

Source: FAO (2014), *The State of World Fisheries and Aquaculture*, Food and Agriculture Publications, Rome, www.fao.org/publications/card/en/c/097d8007-49a4-4d65-88cd-fcaf6a969776/.

1.2 Aquaculture

The year 2014 is a milestone year for aquaculture. It is the first year where human consumption of fish products from aquaculture exceeds that from capture fisheries. However, because of the amount of fish used as feed, capture fisheries production will still exceed aquaculture for some time.

In comparison with the flat trend of marine capture fisheries, aquaculture has been consistently the fastest growing of all food commodities (Figure 1.3). The growing importance of aquaculture has increased the amount of attention paid to the sector by policy makers. The European Union's "Blue Growth" initiative stressed the possibility that EU aquaculture can fill the gap between domestic production and consumption in the European Union, where more than half of fish consumed are imported (European Commission, 2013).

World aquaculture production is also centred in Asian countries, especially by China, India, Viet Nam, Indonesia, and Bangladesh. These five countries account for 80% of global aquaculture production. The major OECD aquaculture producers are Norway, Chile, Japan, Korea, and the United States, who together account for 6% of global production by volume (Table 1.2).

Overall, aquaculture production has grown at an annual 8.6% rate from 1983 to 2012. China alone saw 11.5% annual growth during the same period. In 2012, China produced 23 times as much as it did in 1982 and accounts for 62% of the total production volume in 2012, double its share two decades ago (Figure 1.3).

The growth rate of aquaculture in China in 2012 was 6.4%, less than the world average of 7.5% and less than the OECD growth rate of 7.0%. Compared to the explosive growth of around 25% recorded during the early 1990s, China's growth is slowing. On the other hand, while OECD countries have lost their once-dominant position, growth in the OECD region has been accelerating. Many OECD countries are expected to increase their volume of aquaculture based on strategic plans or guidelines, e.g. National Strategic Plan for Federal Aquaculture Research of the United States, and Strategic Guidelines of the sustainable development of aquaculture of the European Union.

Among OECD countries, Norway is the largest aquaculture producer by volume. Chile is the highest by value with a 23% share in OECD total value, followed by Norway, Japan, Korea and United States. These five countries account for 67% of the OECD total production volume and 68% of the value (Figure 1.4).

Figure 1.3. China dominates global aquaculture but is less dominant in trade

World, OECD and China aquaculture production, 1982-2012

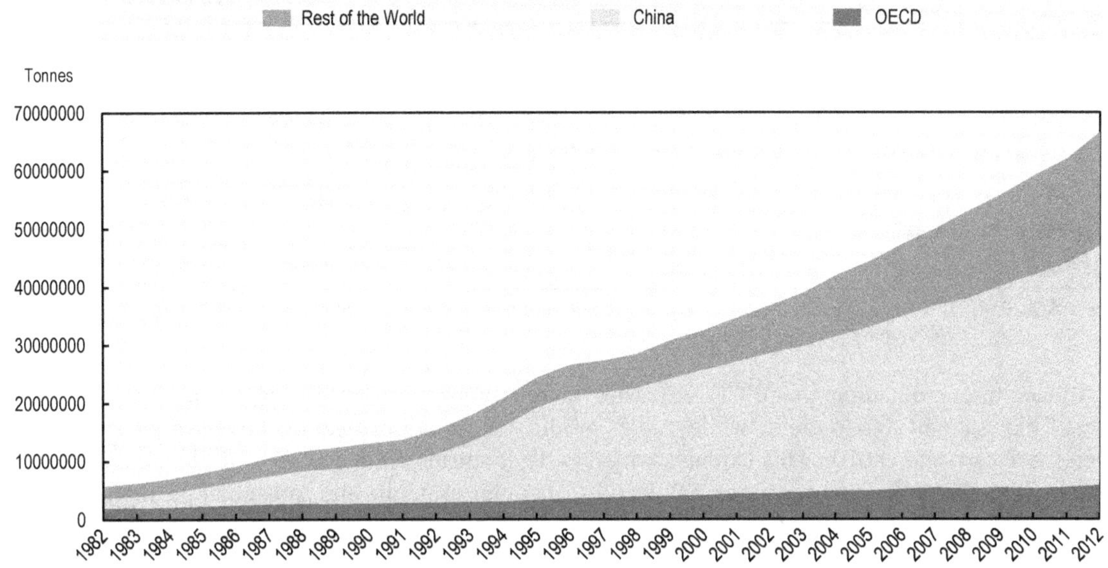

Source: OECD (2015), *Fisheries and aquaculture, OECD Agriculture statistics (Database)* and FAO FIGIS. FishStat (Database), dx.doi.org/10.1787/data-00285-en and www.fao.org/fishery/statistics/global-production/query/en.

Table 1.2. Asian countries extend their lead in aquaculture production

Major producers including top five OECD countries (tonnes)

2012 Ranking	Country	2003	2011	2012	Variation 2003-2012	Variation 2011-2012
1	China	25 083 253	38 621 269	41 108 306	63.9%	6.4%
2	India	2 315 771	3 673 082	4 209 415	81.8%	14.6%
3	Viet Nam	937 502	2 845 600	3 085 500	229.1%	8.4%
4	Indonesia	996 659	2 718 421	3 067 660	207.8%	12.8%
5	Bangladesh	856 956	1 523 759	1 726 066	101.4%	13.3%
6	Norway	584 423	1 143 820	1 321 119	126.1%	15.5%
7	Thailand	1 064 407	1 201 455	1 233 877	15.9%	2.7%
8	Chile	567 259	954 845	1 071 421	88.9%	12.2%
9	Egypt	445 181	9 886 820	1 017 738	128.6%	3.1%
10	Myanmar	252 010	816 820	885 169	251.2%	8.4%
11	Philippines	459 615	767 287	790 894	72.1%	3.1%
12	Brazil	273 268	629 609	707 461	158.9%	12.4%
13	Japan	824 057	556 761	633 047	-23.2%	13.7%
14	Korea	387 791	507 052	484 404	24.9%	-4.5%
15	United States	545 971	397 292	420 024	-23.1%	5.7%

continued

Table 1.2. Asian countries extend their lead in aquaculture production (*cont.*)

2012 Ranking	Country	2003	2011	2012	Variation 2003-2012	Variation 2011-2012
Total 15 countries		35 594 123	57 343 892	61 762 101	73.5%	7.7%
Top 15' share in world		91.4%	92.5%	92.7%	-	-
OECD-top 5		2 909 501	3 559 770	3 930 006	35.1%	10.4%
OECD-5' share in world (OECD)		7.5% (61.7%)	5.7% (64.6%)	5.9% (66.7%)	-	-
OECD-34		4 717 344	5 509 565	5 893 720	24.9%	7.0%
OECD-34's share in world		12.1%	8.9%	8.8%	-	-
World Total		38 915 699	62 011 524	66 633 253	71.2%	7.5%

Source: FAO(2014), *The State of World Fisheries and Aquaculture*, Food and Agriculture Publications, Rome, http://www.fao.org/publications/card/en/c/097d8007-49a4-4d65-88cd-fcaf6a969776/.

Aquaculture production in OECD countries is relatively concentrated in higher value marine species like salmon and oysters, while Asian production is dominated by lower-value freshwater species like carp and catfish. This explains why OECD countries produced 19% of global aquaculture value in 2012, but only 9% by volume. Of the OECD countries, Chile and Japan had the highest value products from aquaculture (Figure 1.5).

OECD countries have made consistent progress in increasing the value of aquaculture products. In the last ten years, OECD production grew by 25% but more than doubled in value. Most OECD countries recorded increased value, even when the volume of production declined (Figure 1.4). The major OECD aquaculture producers, Norway, Chile, Spain, Korea, Japan, and the United States, increased their overall value of production by 119.5% during this period.

Figure 1.4. There is a shift to higher-value aquaculture products in OECD countries

Change in volume and value of aquaculture, 2003-2012 (%)

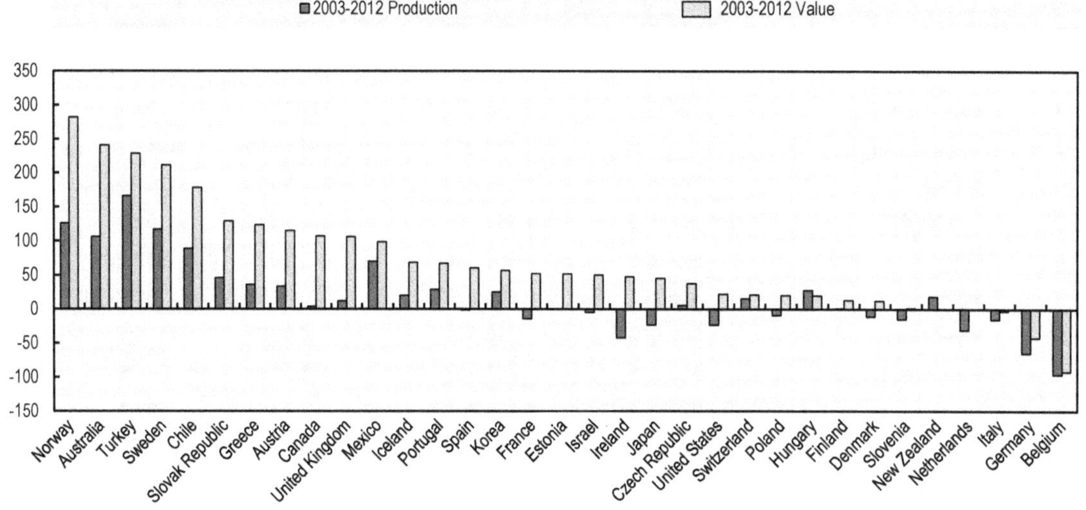

Source: OECD (2015), "Fisheries and aquaculture", OECD Agriculture statistics (Database) and FAO FIGIS FishStat (Database), dx.doi.org/10.1787/data-00285-en and www.fao.org/fishery/statistics/global-commodities-production/en.

Figure 1.5. Aquaculture production is concentrated among a few players

Sources of aquaculture production volume and value, 2012

Volume

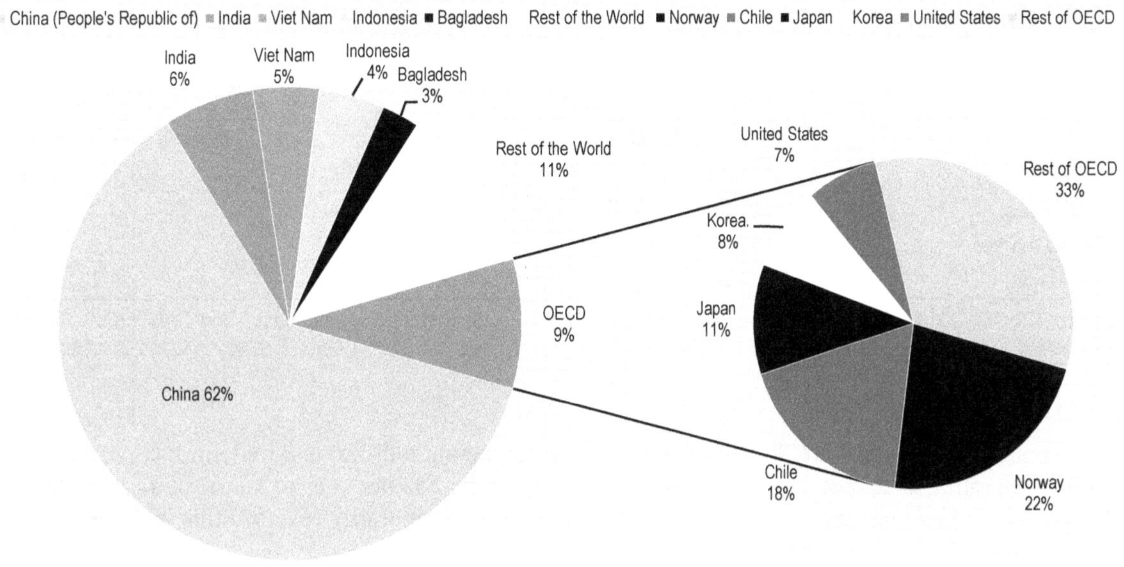

Value

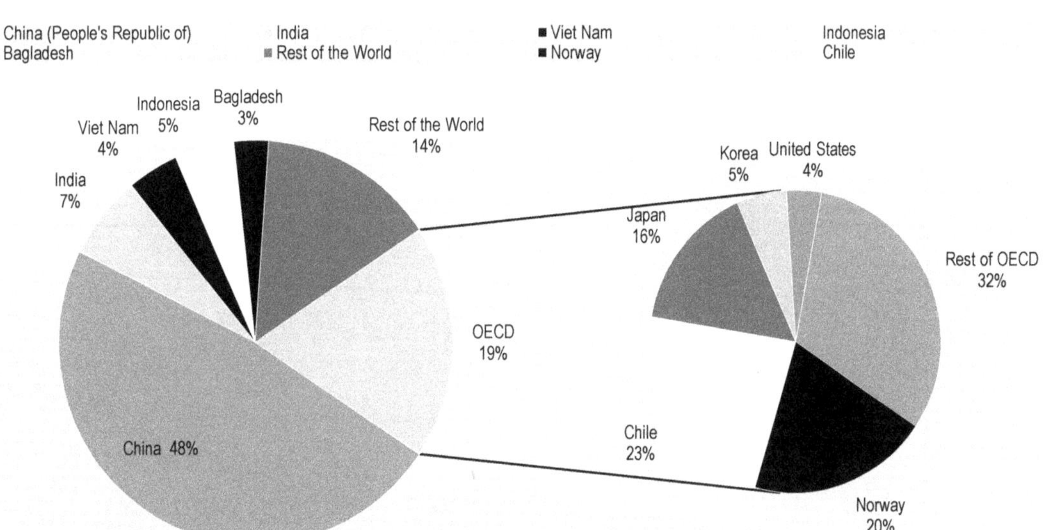

Source: OECD (2015), "Fisheries and aquaculture", OECD Agriculture statistics (Database) and FAO FIGIS FishStat (Database), dx.doi.org/10.1787/data-00285-en and www.fao.org/fishery/statistics/global-commodities-production/en.

Turkey in particular, among OECD countries, has experienced impressive growth, recording a 166% growth in production and 228% increase in aquaculture value in the last ten years. Turkey started aquaculture production in 1986 with 3 000 tonnes, but reached 212 000 tonnes by 2012, producing Bluefin tuna, common sea bream, sharp snout sea bream, meagre and rainbow trout.

The most recent year-on-year data show a slowing of OECD aquaculture value growth rates (Table 1.3). However, it is posited as a temporary phenomenon. For example, the price of salmon was at lowest level as around EUR 3/kg from August 2011 to December 2012. Then, from April 2013, the price recovered at EUR 6.5/kg level. At the end of the year 2014, the price was moving around EUR 5/kg.

Table 1.3. Developments in OECD aquaculture, volume and value (%)

	Volume		Value	
	2003-12	2011-12	2003-12	2011-12
OECD total	24.9	7.0	104.3	-1.7
OECD top five	35.1	10.4	119.5	-2.3
World total	71.2	7.5	154.9	5.8

Source: OECD (2015), "Fisheries and aquaculture", OECD Agriculture statistics (Database) and FAO FIGIS FishStat (Database), dx.doi.org/10.1787/data-00285-en and www.fao.org/fishery/statistics/global-commodities-production/en.

1.3 Trade

Fish and fish products are the most highly traded food commodities, and trade in fish products has been accelerating. Between 2003 and 2012, imports grew by 8% per year in value, more than double the rate of the previous decade, although the financial crisis temporarily reversed this in 2009 and this trend flattened in 2012 (Figure 1.6).

This growth is driven by emerging economies like Indonesia, Viet Nam, Thailand, Brazil, the Russian Federation and Egypt. These countries saw trade growth during last decade of between 190% (Thailand) to 610% (Egypt). During the same period, OECD value of trade grew by 66.8%.

Figure 1.6. Rapid growth in trade, but has a peak been reached?

World and OECD imports of marine products

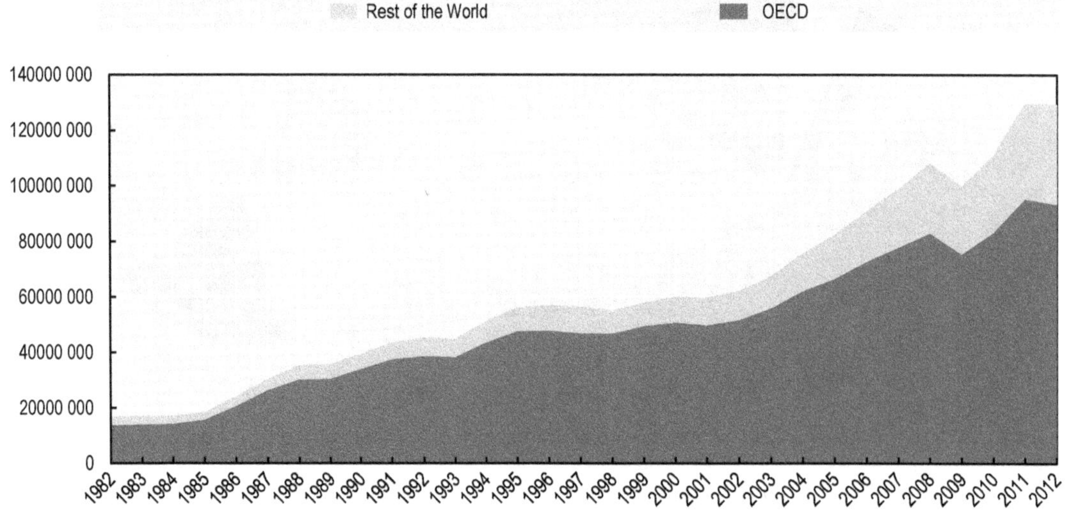

Source: OECD (2015), "Fisheries and aquaculture", OECD Agriculture statistics (Database) and FAO FIGIS FishStat (Database), dx.doi.org/10.1787/data-00285-en and www.fao.org/fishery/statistics/global-commodities-production/en.

Table 1.4. Top ten and OECD top seven exporters of fish and fishery products (USD millions)

2012 Ranking	Country	2003	2011	2012	Variation 2003-2012	Variation 2011-2012
1	China	5 243	16 959	18 211	247.3%	7.4%
2	Norway	3 624	9 456	8 895	145.4%	-5.9%
3	Thailand	3 929	8 141	8 078	105.6%	-0.8%
4	Viet Nam	2 199	6 241	6 277	185.4%	0.6%
5	United States	3 398	5 788	5 753	69.3%	-0.6%
6	Chile	2 134	4 504	4 337	103.2%	-3.7%
7	Canada	3 330	4 198	4 213	27.7%	0.3%
8	Denmark	3 221	4 482	4 147	29.1%	-7.5%
9	Spain	2 224	4 185	3 951	77.7%	-5.6%
10	Netherlands	2 182	3 549	3 878	77.7%	9.2%
Top ten subtotal		31 451	67 509	67 743	115.4%	0.3%
Top ten's share in the world		49.0%	52.1%	52.4%		
OECD-top 7		20 078	36 166	35 176	75.2%	-2.7%
OECD-7' share in world(OECD)		31.3%	27.9%	27.2%		
OECD-34		33 769	62 976	60 743	79.9%	-3.5%
OECD-34's share in world		52.7%	48.6%	47.0%		
World total		64 122	129 594	129 298	101.6%	-0.2%

Source: FAO (2014), *The State of World Fisheries and Aquaculture*, www.fao.org/publications/card/en/c/097d8007-49a4-4d65-88cd-fcaf6a969776/.

Despite the OECD's moderate share of total production, global trade of fisheries products are still dominated by OECD countries (Figure 1.7). OECD countries accounted for around 72% of all imports in 2012, down from 83% in 2003. In terms of exports, OECD countries accounted for 47% of the total in 2012, compared with 52.7% in 2003. The largest consumers of fish products are the European Union, the United States and Japan.

In 2012, seven of the ten top trading nations for fish products were OECD countries (Norway, the United States, Chile, Canada, Denmark, Spain and the Netherlands) were listed in the top ten list of fisheries exporters (Table 1.4). Asian countries dominate production, but typically, and especially in the case of China, consume the majority of production domestically. By contrast, Norway and Chile are export-oriented and consume only a small part of their domestic production. While China's imports and exports are small relative to production, it is still among the top importers and exporters globally.

The import dominance of OECD countries enables big importers like the United States and the European Union to set standards for imported products that influence production and processing. For example, controls are implemented to prevent the import of illegal, unreported, and unregulated (IUU) fish products, or to maintain certain sanitary standards. Also, exporters must meet certain product specifications in some cases. For example, in the United States shrimp market Thailand used to be the leading supplier. However, in the first half of 2014, exports from Viet Nam replaced much of the supply from Thailand because of ability to provide consistent sizes and volumes (Seafood International, 2014).

Table 1.5. Top ten and OECD top eight importers of fish and fishery products (USD millions)

2012 Ranking	Country	2003	2011	2012	Variation 2003-2012	Variation 2011-2012
1	Japan	12 395	17 340	17 988	45.1%	3.7%
2	United States	11 653	17 466	17 561	50.7%	0.5%
3	China	2 388	7 572	7 441	211.5%	-1.7%
4	Spain	4 904	7 309	6 487	32.3%	-11.2%
5	France	3 771	6 567	6 040	60.2%	-8.0%
6	Italy	3 558	6 211	5 563	56.3%	-10.4%
7	Germany	2 635	5 513	5 305	101.3%	-3.8%
8	United Kingdom	2 507	4 257	4 252	69.6%	-0.1%
9	Korea	1 950	3 935	3 736	91.5%	-5.0%
10	Hong Kong, China	1 752	3 513	3 663	109.0%	4.3%
Top ten subtotal		47 518	79 687	78 041	64.2%	-2.1%
Top Ten's share in the world		70.5%	61.4%	60.3%		
OECD-top 7		27 829	47 465	46 123	65.7%	-2.8%
OECD-7' share in world(OECD)		41.3%	36.6%	35.6%		
OECD-34		55 936	95 265	93 295	66.8.%	-2.1%
OECD-34's share in world		83.0%	73.4%	72.1%		
World total		67 390	129 805	129 466	92.1%	-0.3%

Source: FAO (2014), *The State of World Fisheries and Aquaculture*, Food and Agriculture Publications, Rome, www.fao.org/publications/card/en/c/097d8007-49a4-4d65-88cd-fcaf6a969776/.

Figure 1.7. OECD countries are the largest importers and exporters of fish products

Share of OECD in world trade (%)

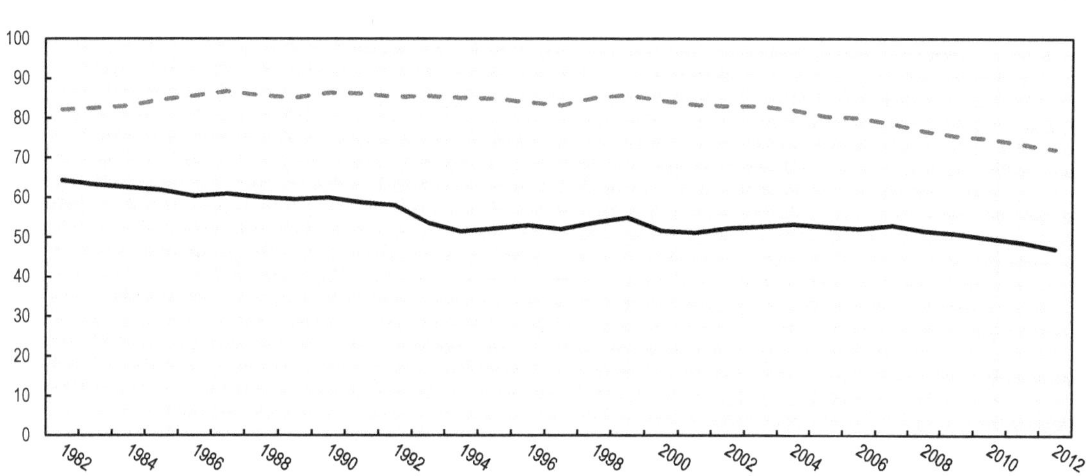

Source: OECD (2015), Fisheries and aquaculture, OECD Agriculture statistics and FAO FIGIS FishStat (Database), dx.doi.org/10.1787/data-00285-en and www.fao.org/fishery/statistics/global-commodities-production/en.

Emerging fisheries producers such as Indonesia and Thailand saw stronger growth in imports than exports. Those two countries recorded import growth of 375.6%, and 189% respectively compared to 131.6% and 105.6% growth of exports from 2003 to 2012 period (Table 1.6).

China has rapidly become an important trader of fisheries products, but it is not certain whether this rate of growth can be sustained. The value of exports from China to the world increased tenfold from 1992 to 2012, an annual rate of 13.8% (Figure 1.8). In the last decade it increased by 247%; a 15.3% annual growth rate. Between 2011 and 2012, the growth rate slowed to 7.3%. Whether this is due to global market conditions or China reaching the limit of its production capacity will be made clear in the coming years.

Table 1.6. Major fisheries exporters' growth rate, export and import, value

Country	Imports (%)		Exports (%)	
	2003-12	2011-12	2003-12	2011-12
Chile	383.2	2.6	103.2	-3.7
Viet Nam	457.1	15.5	185.4	0.6
Thailand	189	14.4	105.6	-0.8
Canada	88.6	1.6	27.7	0.3
Netherlands	100.9	3.9	77.7	9.2
Denmark	49.6	-3.1	29.1	-7.5
Norway	142.1	1.7	145.4	-5.9
United States	50.7	0.5	69.3	-0.6
China	211.5	-1.7	247.3	7.4
Spain	32.3	-11.2	77.7	-5.6

Source: OECD(2015) Fisheries and Aquaculture, OECD Aquaculture Statistics (Database) and FAO FIGIS FishStat (Database), dx.doi.org/10.1787/data-00285-en and www.fao.org/fishery/statistics/global-commodities-production/en.

Figure 1.8. China shows strong growth in imports and exports as trade gap widens

Chinese exports and imports, tonnes 1992-2012

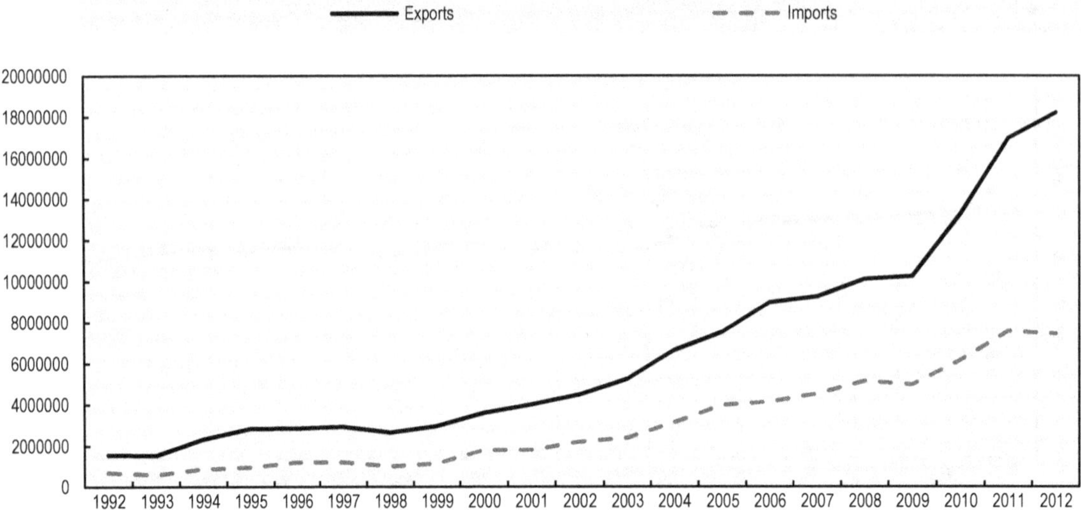

Source: OECD (2015), Fisheries and aquaculture, OECD Agriculture statistics and FAO FIGIS FishStat (Database), dx.doi.org/10.1787/data-00285-en and www.fao.org/fishery/statistics/global-commodities-production/en.

Figure 1.9. China's share in world imports and exports (%)

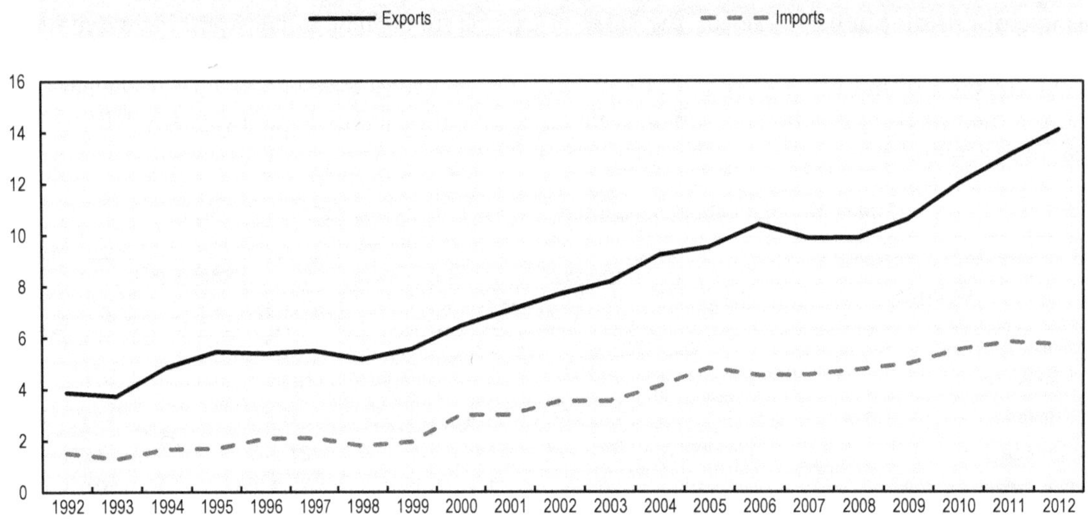

Source: OECD (2015), Fisheries and aquaculture, OECD Agriculture statistics and FAO FIGIS FishStat (Database). dx.doi.org/10.1787/data-00285-en and www.fao.org/fishery/statistics/global-commodities-production/en.

Economic growth in China has been driving an increase in consumption of fisheries products. Per capita consumption in China's cities rose from 10.34 kg in 2000 to 14.62 kg in 2011 (Fisher and Mutter, 2014). Growth is particularly strong for high value species like salmon.

Trade for re-export clouds the picture for Chinese trade. Many companies ship raw material to China for processing and then re-import the final products. However, the importance of this trade may be decreasing as western fisheries producers are expected to move processing back to Europe and North America, in particular for Atlantic species. This is due to increased production costs in China as well as higher transport costs between Europe and China.

1.4 Outlook

This section reports the results of projections made using the OECD-FAO Aglink–Cosimo model. These results can also be found in the publication *OECD-FAO Agricultural Outlook 2015* (OECD/FAO 2015). This model estimates for future fisheries market outcomes, including production, trade, prices, human consumption, and ratio of fishmeal consumption under the following assumptions:

- The price of meats and feed ingredients will remain high.
- Fishing quota under-fill will be minimal. That is, capture fisheries are constrained by quotas and cannot increase in response to higher fish prices.
- Aquaculture productivity gains will be smaller than in the previous decade.
- New feeding techniques will not prevent increase in the ratio of fish to oilseed meal price.
- The popularity of omega 3 fatty acids and growth in aquaculture production has permanently increased the fish to oilseed oil price ratio.

Under these assumptions, it is expected that total fish production will grow substantially by 2024, entirely driven by growth in aquaculture production. The total volume of fish and seafood production is estimated to be 191 million tonnes by 2024, corresponding to 19% growth from the base period. Aquaculture is estimated to grow 38% over this period while capture fisheries are expected to increase by only 1%. The estimated growth in aquaculture has been revised upward in the last two Outlook projections. However, the annual growth rate is estimated to be only 2.5% between 2014 and 2024, significantly lower than historical levels of 6% or more. This is due to competing uses of the coastal space, and higher costs of fishmeal, fish oil and other feeds. It is expected that average annual growth rate of aquaculture will fall from 18% in the 1980s to around 2% by the 2020s. This is at least in part due to the maturing of the sector, as growth is calculated from an increasingly large base.

China's share of aquaculture production is projected to be stable at more than 60% of the global total, with a 37% share of world fishmeal consumption by 2024. In case of capture fisheries, the share of China is expected to decline from 18% in 2014 to 16% in 2024.

The share of aquaculture in human consumption is estimated to reach 56% by 2024, 96 million tonnes from aquaculture and 79 million tonnes from capture fisheries. Among 191 million tonnes of total fisheries production in 2024, about 90% (172 million tonnes) is estimated to be used for direct human consumption (Figure 1.10).

Both nominal and real prices of fish and fish products are generally expected to rise by 2024 (Figure 1.11). In general, it is expected that production costs will increase faster than the price of products. Many factors are contributing to an upward trend in the price of fish. These are income and population growth, relatively stable capture, increasing feed cost and meat prices, lower productivity gain in aquaculture and high crude oil price.

Figure 1.10. Aquaculture has surpassed capture fisheries as main source of human consumption

Consumption of fisheries products (million tonnes)

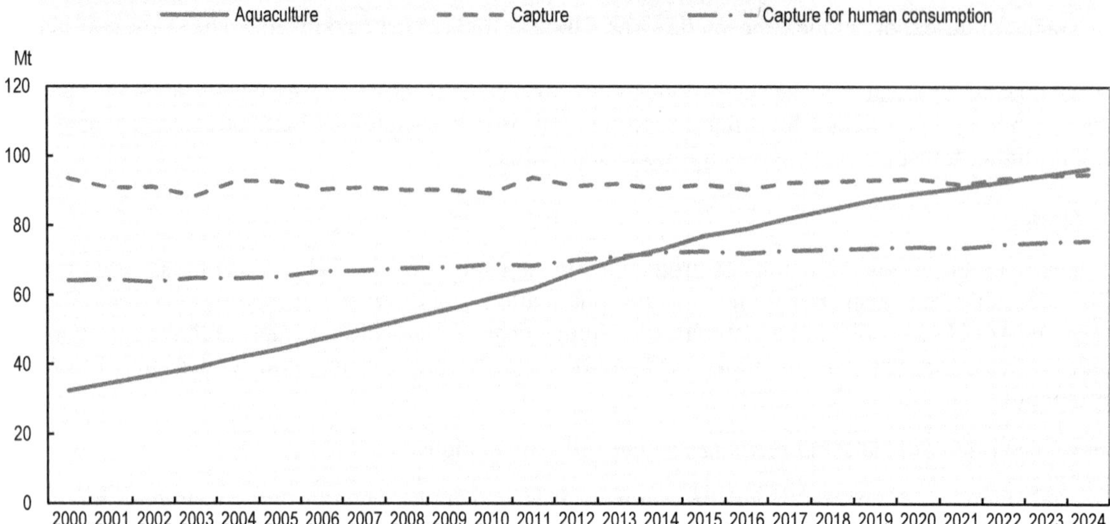

Source: OECD (2015), Aquaculture and capture fisheries, in *OECD-FAO Agricultural Outlook 2015*, dx.doi.org/10.1787/agr_outlook-2015-graph56-en.

Figure 1.11. Rising price trends for most products

Real prices (US GDP deflator) left panel. Nominal prices right panel

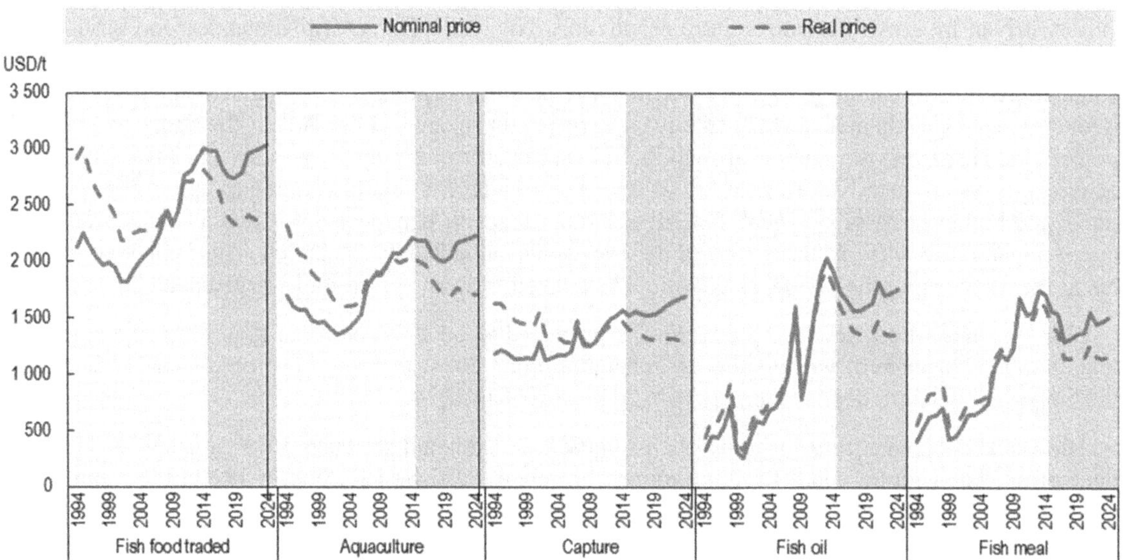

Source: OECD (2015), "World fish prices", in *OECD-FAO Agricultural Outlook 2015*, doi.org/10.1787/agr_outlook-2015-graph100-en.

Figure 1.12. Broad growth in consumption worldwide

Increase in fish consumption by region to 2023

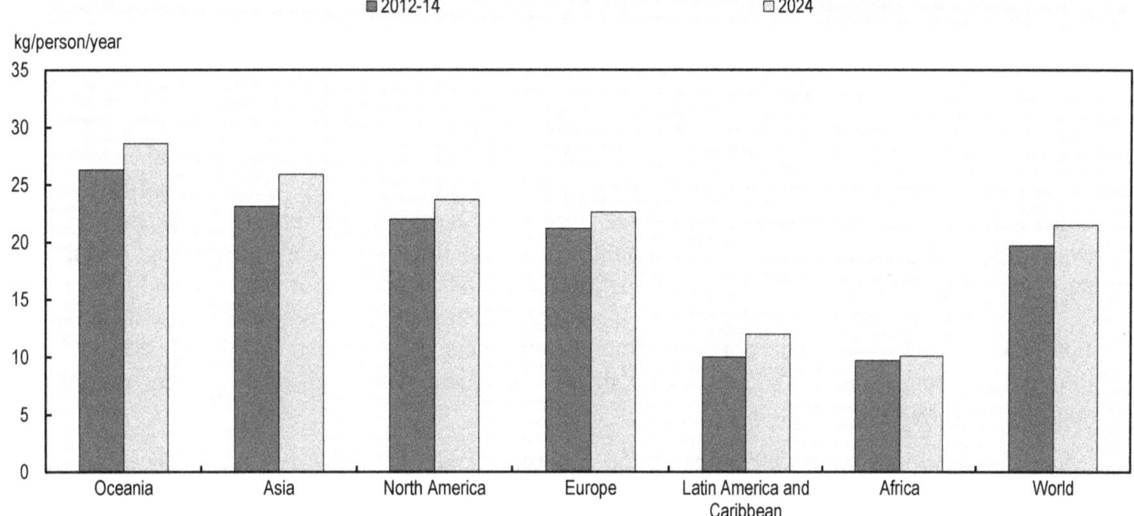

Source: OECD (2015), "Apparent per capita fish consumption by continent", in *OECD-FAO Agricultural Outlook 2015*, dx.doi.org/10.1787/agr_outlook-2015-graph102-en.

An increase in per-capita fish consumption is expected in all world regions (Figure 1.12). Consumption is expected to grow by 30 million tonnes by 2024, 81% of which will be in Asia followed by Africa (9%), Latin America, Caribbean (6%), North America (3%), and Oceania (1%).

The ratio of fishmeal consumption to animal feed will continue to decline (Figure 1.13). The higher cost of fishmeal is driving reductions in the inclusion rate as aquaculture producers seek lower cost alternatives. Productivity gains in aquaculture will also be smaller due to environmental regulations, density limits to control animal diseases and an exhaustion of optimal production locations.

Figure 1.13. Fishmeal will form only a small part of animal feed rations

Fishmeal inclusion rate in aquaculture feeds, %

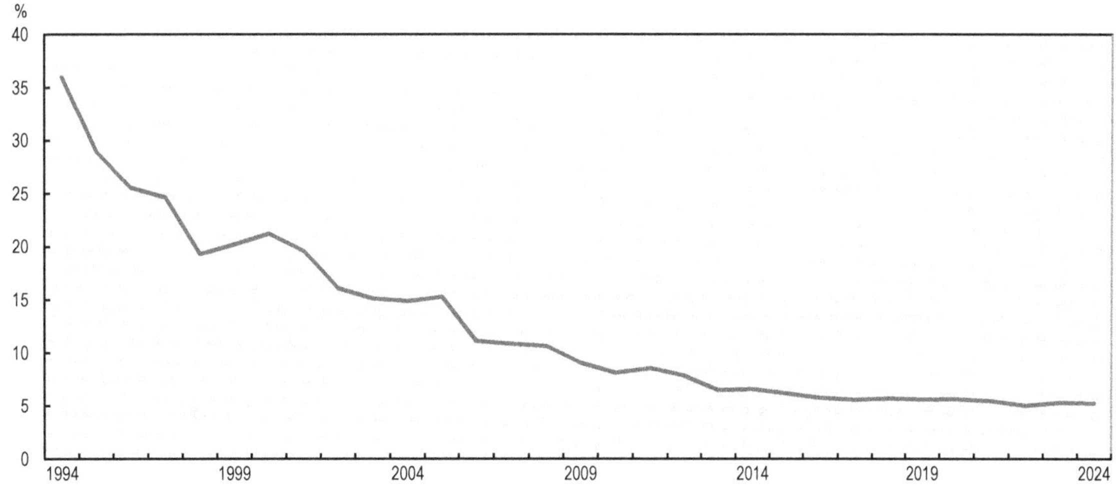

Source: OECD/FAO (2015), "OECD-FAO Agricultural Outlook", *OECD Agriculture Statistics* (database), dx.doi.org/10.1787/agr-outl-data-en.

In 2024, fishmeal production will reach 5.1 Mt (product weight) and fish oil 1 million Mt (product weight), up 9% and 13%, respectively, compared to the averages of 2012-14. Fishmeal and fish oil can be produced from whole fish, fish remains or other fish by-products such as heads, tails, bones and other processing by-products.

Fishmeal and fish oil from whole fish are mainly produced from capture fisheries. It is expected that the percentage of harvested fish converted to fishmeal will remain stable at around 16%. That share will be slightly smaller in years of El Niño, which is associated with reduced anchoveta catches, a species mainly used for reduction into fishmeal/fish oil. The growing demand for human consumption for species previously used for reduction is already driving up prices and shifting supply for some pelagic species in Norway and Iceland. In part this is driven by improved management which allows fishers to maintain a higher product quality for fish, and thereby entering consumption markets rather than selling their catch for reduction. Sustained demand and prices for fishmeal and fish oil will make production of fishmeal and fish oil from fish waste and by-products of processing more economical. In 2024, fishmeal obtained from by-products is estimated to reach 28% of total production, up from 26% in 2012-14. For fish oil, this share will reach 43% of total production, compared with 40% in 2012-14.

After a sharp decrease in 2009, due to the economic and financial crisis, trade has returned to positive growth but will be at a structurally lower rate because of the increasing transportation cost resulting from the high fuel prices and slower growth in aquaculture production (Figure 1.14). Total fish and fish products will continue to be traded in significant quantities, driven by sustained demand, innovations and improvements in processing, preservation, packaging, transport and logistics. About 31% of total fish production (excluding European Union intra-trade) will be exported by 2024. The fisheries supply chain will remain complex as outsourcing of processing to developing countries will continue to be a major factor.

Figure 1.14. Developing country exports have grown strongly

Trade in fish, millions of tonnes

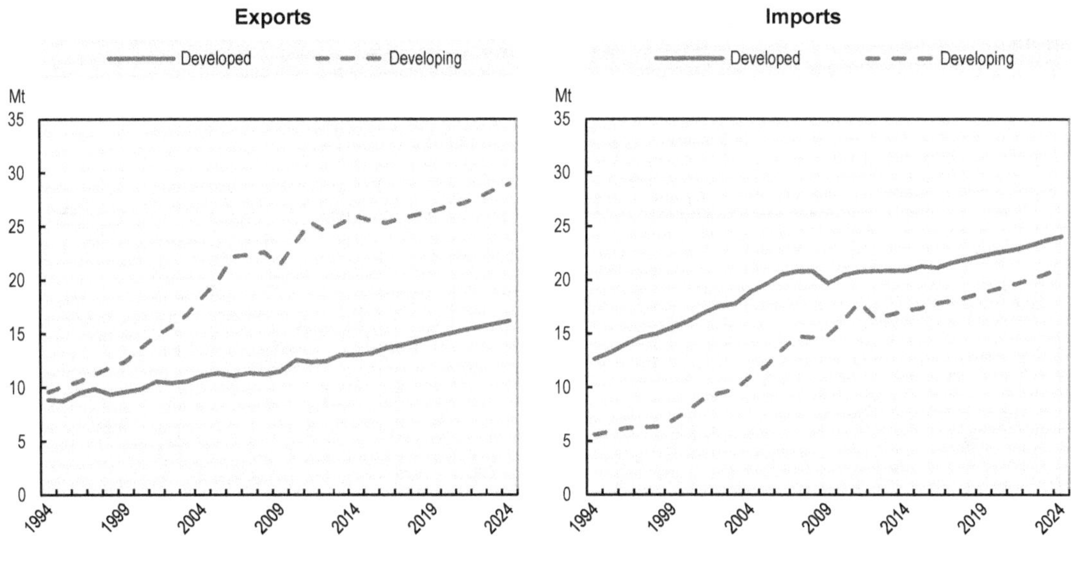

Source: OECD/FAO (2015), "OECD-FAO Agricultural Outlook", OECD Agriculture Statistics (database), dx.doi.org/10.1787/agr-outl-data-en.

World trade of fish is projected to increase by 14% in volume from 2014 to 2024. The share of developed countries in world fish imports will decline from 69% currently to 53% in 2024, as domestic demand in developing countries matures. Developing countries are expected to represent 64% of world exports in 2024, with the majority of world fish exports for human consumption originating from Asia. Owing to their importance in aquaculture production, developing countries will continue to be the main importers of fish meal. China alone should represent 62% of world aquaculture production by 2024 and consume 43% of world fishmeal production.

A note on fish meal and oil: A challenge for aquaculture development

Aquaculture and capture fisheries currently provide almost the same amount of seafood for human consumption. However, a considerable amount of capture fisheries landings are used for reduction to fishmeal and fish oil rather than human consumption and this has supported the growth in aquaculture production. If aquaculture is to continue to grow, it will have to do so in the context of limited potential for increased supply of fishmeal and oil from capture fisheries.

There are three main categories of sources of raw material for reduction; (1) industrial forage including menhaden and sandeel; (2) food grade forage fish like anchovy, capelin and whiting; and (3) prime food fish such as herring, sardines, and mackerel. In the case of first category, it is virtually impossible to sell the fish for other purposes because of their small size and bony characteristics. For the second category, these fish have limited demand for direct human consumption. However, when the third category fish is used for reduction, it is mainly because of poor fisheries management which leads fishers to a race to fish, reducing quality and intermittently flooding markets.

In the Norwegian and Chilean cases, it has been shown that the introduction of more efficient management systems, such as individual vessel quotas (IVQs), can contribute to better use of fisheries resources. In the 1960s Norwegian herring were mostly used to make fishmeal. After the introduction of IVQs for the purse seine fleet and in 2005 for coastal vessels, the proportion of the landings reduced as fishmeal decreased drastically. In 2009, the share used for fishmeal was around 10% compared with 90% in the 1960s. When Chile introduced IVQs for Jack mackerel in 2001 the share of landings used for fish meal also decreased from 90% in 1996 to 10% in 2012. While this is better for fishers and consumers, reduced fishmeal supply is a challenge for aquaculture.

In Norway virtually all, trimmings of pelagic fish are recovered for fishmeal and oil production. In the case of the whitefish sector, the percentage of the utilisation of trimmings is around 40%. The difference is due to the fact that whitefish are harvested by small coastal vessels scattered along the Norwegian coast, such that the economies of scale to actually use the trimmings do not exist, leading to trimmings being dumped at sea.

The Norwegian salmon aquaculture industry shows the importance of scale. In the 1980s, most trimmings were dumped. However, as quantities reached commercial scale several salmon companies set up facilities to collect and transport trimmings to processing plants. Today the Norwegian salmon industry collects and re-uses nearly all of its trimmings.

The main use of fishmeal and fish oil was historically for feed for pigs and chicken. In the 1960s this accounted for 98% of total fishmeal use. By 2010 the proportion of feed for these animals was only 25% while the share of fishmeal going to aquaculture reached 73%. The story for fish oil is similar, though direct human consumption also has increased tremendously during the last decade due to the emergence of Omega-3 supplements.

As prices have been going up for aquaculture feed components, feed producers have worked hard to develop alternative sources. While dramatic progress has been made in replacing fishmeal and oil with land-based alternatives such as soy meal, some species like salmon and sea bass remain dependent on a certain level of fish oil. For these species, the competing uses of fish oil may constrain future production.

References

Fisher, E. and R. Mutter (2014), “China under the microscope”, *Fishing News International*, Vol. 53, Issue 1, fishingnewsinternational.com/?s=China+under+the+microscope.

Food and Agriculture Organization of the United Nations (FAO) (2014), *The State of World Fisheries and Aquaculture*, Food and Agriculture Publications, Rome, www.fao.org/publications/card/en/c/097d8007-49a4-4d65-88cd-fcaf6a969776/.

Food and Agriculture Organization of the United Nations (FAO). FIGIS. FishStat (Database). www.fao.org/fishery/statistics/global-production/query/en.

OECD/FAO (2015), *OECD-FAO Agricultural Outlook 2015*, OECD Publishing, Paris. dx.doi.org/10.1787/agr_outlook-2015-en.

OECD (2015) Fisheries and aquaculture, OECD Agriculture statistics (Database) dx.doi.org/10.1787/data-00285-en.

Chapter 2

Fishery policy developments in OECD countries

This chapter offers an overview of policy developments in OECD countries as well as the activities of the OECD Committee for Fisheries (COFI). Many OECD countiers and partner economies are undertaking important structural and policy reform in their fisheries sector, and new governance structures and management instruments are being put in place. In particular, the recent reforms of the EU Common Fisheries Policy will bring significant changes in fisheries management in the EU.

2.1 Major activities of OECD Fisheries Committee

Green growth in fisheries and aquaculture

In 2010, the OECD Ministerial Council tasked the OECD to identify green policy options and market approaches that would encourage green growth including mitigating the food system's contribution to climate change. A year later, the Green Growth Strategy (GGS) Synthesis report was adopted by OECD Ministers. This work provided the conceptual and policy base for countries to take the green growth agenda forward (OECD, 2011).

The OECD GGS is based on the idea that it is both possible and necessary to achieve sustained economic growth while reducing the human impact on the environment. This is particularly relevant for fisheries, which is more dependent than most economic sectors on environmental resources, both fish stocks and the health of the marine ecosystem. Evidence suggests that the fisheries and aquaculture sectors are able to deliver growth in the short and long term, so long as they are in an enabling policy environment.

There are a number of ways in which fisheries can increase its economic and social contribution while respecting the natural limits of the ecosystem. First and foremost, the depleted state of many fisheries means that rebuilding these stocks can lead to higher future harvests. The value to the consumer of fisheries products will also continue to increase as new products and new markets emerge notably through innovation. Processors and other value-chain participants will find new uses for what is currently wasted. Fishers will continue to reduce their costs, improve their efficiency and productivity and earn a greater return from their efforts.

The fisheries and aquaculture sectors depends on effective public institutions and good governance. Good institutions tend to reflect three characteristics: good availability of information about the actions engaged in or supervised by the entity, transparency in the decision-making process, and accountability in decisions made and enforcement measures taken. National development plans, institutional innovation, certification, spatial planning and public-private partnerships have all been identified as ways to improve the prospects of aquaculture. Greening the sector also leads to lower production costs as energy costs tend to be better managed, costs associated with disease outbreaks, waste and escapees are reduced, while innovations lead to better management of production costs. In essence, when risks associated with aquaculture are addressed early on and when innovation is embraced, costs of production tend to be lower.

Both the fisheries and aquaculture sectors would benefit from a multi-sector approach that minimises negative impacts on other users. The overall use of space and water resources (whether marine or on land) and the potential for conflicts between different user groups is an important challenge which needs to be addressed nationally and internationally. Governance mechanisms that take into account the contribution of different sectors to the ocean economy can help ensure that stakeholders co-operate with multiple government agencies across a wide area of interests, and across countries sharing marine resources.

Green growth is a process of continual improvement. Finding more efficient ways to do business and to derive value from fish resources is an on-going exercise. Sound internal feedback mechanisms and sharing of experiences and best practices between countries is a good way to find pathways to improvement. Developing countries in particular can benefit from international cooperation.

Improving the Government Financial Transfers Database

The OECD collects a unique set of data on budgetary policies in participating states. Called the Government Financial Transfers (GFT) database, this information serves the core mission of the OECD to provide internationally-comparable data to support evidence-based policy analysis.

To further improve the quality and utility of the GFT database, the OECD Secretariat under the guidance of the COFI has undertaken to review and revise the content of this database as well as its organisation and presentation. The new database is intended to provide information on financial transfers in much greater detail than ever before. The process of collecting data for the GFT database will become more systematic and reliable, with improved capacity for data validation by the Secretariat. This will help ensure a high level of data quality and reliability.

The new GFT exercise is intended to capture the economically-relevant features of fisheries policies in a way that supports understanding and analysis of the way policies operate. This means not only a classification system based on the way programs are implemented but also a complementary system of "labels" that capture additional details about how support programmes work. These details concern the eligibility conditions under which the associated transfers are provided to fishers. The objective to have information on policies' implementations is to see if some differences exist between countries, to share experiences and identify the best practices for policy makers.

Non-budgetary transfers such as tax concessions and exemptions are an important form of support in many OECD countries. Including these policies in the GFT database is crucial to understanding the whole policy set. The OECD has recently undertaken a study of fuel tax concessions granted by countries to their fishing sectors, and the approach developed in that study will be improved and systematically included in the GFT database.

The GFT database will be part of a broader improvement in analytical capacity, where GFT data underlies the development of indicators of performance in the fisheries sector, potentially including measures of impact on welfare, incomes, biodiversity, fishing effort and more.

Joint OECD-DAC-FAO-WB workshop on Policy Coherence for Development in Fisheries and Aquaculture

The Fishing for Development Workshop investigated many questions central to developing sustainable fisheries and aquaculture policies in developing, emerging and developed countries alike. How can fisheries and aquaculture best contribute to economic development and food security, and particularly through more coherent policy making? How can the synergies between fisheries and aquaculture and development assistance policies be exploited to support sustainable development and contribute to green growth? How can governments' capacity to better balance sometimes divergent domestic fisheries and aquaculture objectives with broader development cooperation goals be developed in ways that include avoiding or minimising potentially detrimental effects of such policies in developing countries (OECD, 2006)?

Organised by the COFI, the OECD Development Assistance Committee (DAC), FAO and WB, and taking place in Paris in April 2014, this meeting initiated a dialogue between the fisheries and the development policy communities from OECD countries and partner economies on key issues of shared interest. OECD/FAO (2015) highlights the main conclusions of the meeting and includes the background papers that were originally prepared to provide context for the issues addressed. It identifies questions for a future work agenda on policy coherence in fisheries and aquaculture, and makes evident the strong need for further dialogue between the fisheries and development communities at the global and regional scale.

Fisheries and aquaculture contribute significantly to reducing poverty and food insecurity worldwide. It is estimated that this sector, including secondary activities, provides livelihoods for over 600 million people, possibly even more than 800 million. Fish is also an important source of protein, fatty acids and micronutrients that are fundamental to the development of humans, especially in the poorest parts of the world. At the macro-level, fisheries and aquaculture have important economic multiplier and spill over effects and can generate significant government revenues and foreign currency when sustainably managed.

Sustaining the capacity of world fisheries and aquaculture to provide food and jobs requires sensible and effective fish stock management and ecosystem preservation. As fish stocks are often shared between countries, this can only be achieved through regional and multilateral co-operation. In addition, given that oceans function as a single ecosystem, that fish is one of the most traded food commodities, and that fishing activities are mobile, policies that affect fish production, consumption and trade in the OECD and emerging economies are likely to have a significant impact on the development prospects of developing countries. They will also have an impact on the demand for fish produced in developing countries, on global prices for fish products and fish stock and ecosystem sustainability. Concurrently, policies that affect fish production, consumption and trade in developing countries are likely to have an impact on the availability and affordability of fish products in major consumption markets, such as the OECD and emerging economies.

There are many issues where policy coherence is at stake. Four topics which are high on the international fisheries and aquaculture policy agenda were addressed during the meeting: the challenges of rebuilding fish stocks while securing the integrity of ecosystems and the livelihoods that depend on them; the potential for green growth in aquaculture; combatting illegal, unreported and unregulated (IUU) fishing; and the role of regional fisheries management organisations (RFMOs) in the management of high seas fish stocks and in developing co-operation between states that share fish stocks in several EEZs. The conclusions of the meeting and key recommendations are forthcoming in a report to be jointly published with the FAO.

The future of the ocean economy: Exploring the prospects for emerging ocean industries to 2030

Followed closely by the COFI, under the auspices of the International Futures Programme, in 2013 the OECD initiated a project seeking to explore the future prospects of the ocean based industry especially with respect to emerging areas such as ocean-based energy, off-shore oil and gas, seabed mining, marine aquaculture, marine biotechnology, marine tourism, and maritime monitoring. The work seeks to better understand the ocean economy's long-term outlook and future contribution to growth and jobs.

Against the background of increasing demand for food, jobs, energy, raw materials, and economic growth, the ocean industries have a great potential. However, competition among the diverse set of ocean activities, for example aquaculture and ocean energy, suggests that a coordinated and well planned usage of ocean is needed. To tackle this, considerable R&D effort, investment, and coherent policy support are necessary. In this regard, this project aims to provide a forward-looking, cross-sectoral, and multi-disciplinary assessment of challenges and needs.

2.2 The Common Fisheries Policy (CFP) in the European Union

Almost 5 million tonnes of fish are caught annually by EU fleets, with the Danish, Spanish, United Kingdom and French fleets accounting for about half. The European Union adopted in February 2013 a new CFP which came into force on 1 January 2014. The new CFP is a major change from previous versions, first introduced in 1983 and revised every ten years. The new CFP emphasises the need to ensure sustainable fisheries and requires quotas to be set with reference to Maximum Sustainable Yield (MSY) by 2015 in most cases. The new CFP has as its objective rebuilding all fish stocks to MSY levels by 2015, or 2020 at the latest.

The new CFP does away with the wasteful practice of discarding through the introduction of a landing obligation. This is expected to drive more selectivity in fishing gear and practices, and provide more reliable catch data. The landing obligation will be introduced gradually starting in 2015 for pelagic species and between 2016 and 2019 for all commercial fisheries in European waters. Under the landing obligation all catches have to be kept on board, landed and counted against quotas. Undersized fish cannot be marketed for human consumption. Quota management will also become more flexible in its application to facilitate the landing obligation (See http://ec.europa.eu/fisheries/cfp/fishing_rules/discards/index_en.htm).

The CFP is a comprehensive set of regulations dealing with management, international relations, markets and trade, and financing. On the markets and trade side a new provision requires that

Producer Organisations (PO) shall set up and submit plans for the production and marketing of fish. This is mandatory for Producer Organisations to be eligible for certain financial schemes. The framework also sets up common marketing standards with uniform characteristics for certain fish products sold in the European Union. Fish products, seaweed and algae sold to consumers or caterers must bear information related to:

- species' commercial and scientific names
- whether the products were caught at sea, in freshwater, or farmed
- the catch or production area:
  - fish caught at sea: the FAO sub-area or division (NE Atlantic, Mediterranean and Black Sea) or the FAO area (other waters)
  - freshwater fish: the body of water in the EU country or non-EU country of origin
  - farmed fish: EU or non-EU country of final rearing period
- fishing gear used.

The CFP provides for new principles regarding bilateral fisheries agreement ("Sustainable Fisheries Agreements") including provisions limiting access to resources that are scientifically demonstrated to be surplus to the coastal State's own catch capacity. These agreements will now include a clause to protect human rights and gradually increase EU ship-owners' contribution to the access costs. Reforms are also intended to better promote sustainable fishing in the partner country waters by making EU sectoral support more targeted and subject to regular monitoring.

The European Union established a system to prevent, deter, and eliminate illegal, unreported and unregulated fishing (Council Regulation (EC) No 1005/2008). Products may only be sold on the EU market that are certified as legal by the flag state. As of March 2013, the European Union listed Belize, Cambodia, and Guinea-Conakry as countries not cooperating in the fight against IUU fishing. Imports into the European Union of any fisheries products caught by vessels flying flags from these countries were banned. Also, EU vessels will not be allowed to operate in these countries' territorial waters.

2.3 Integrated ocean management

Korea has one of the longest-running integrated ocean management plans. The Ministry of Maritime Affairs and Fisheries (MOMAF) established in the mid-1990s a long-term development strategy for ocean-related matters which balanced environmental and fisheries issues, integrated coastal management and monitoring of fisheries as well as provided a coherent policy for the shipping industry, port construction, and maritime safety. After a five-year period which saw the ocean portfolio split into several units, mounting pressure from fishers' organisations and the shipping industry led to the creation in 2013 of the Korean Ministry of Oceans and Fisheries, with responsibilities to provide a fully integrated approach to all marine issues.

A good illustration of this integrated approach is the monitoring of seaweed aquaculture farms using satellite technology. Previously, collecting information on the use of permits by seaweed farmers was quite difficult. The integrated ocean ministry facilitated the use of satellites to do so and made production forecasting feasible. Thanks to the information gathered over almost ten years, the management of compensations for victims of a major oil spill in 2007 was also greatly improved, particularly for farmers who could not submit exact data about their actual production. Another example is the improvement of safety and prevention through the integrated management of fishing vessels and commercial shipping vessels.

China is also experimenting with unified marine governance in order to protect their ocean interests and develop ocean-related industries. In 2013 four of its five maritime law-enforcement commands were consolidated into the SOA (State Oceanic Authority). The four commands deal with maritime boundaries, fisheries supervision, control of smuggling at sea, illegal activities and environmental surveillance. A high-level co-ordinating body called the National Ocean Committee was also formed, bringing together leadership from multiple ministries to formulate China's ocean development strategy.

Since 2010 the UK's Marine Management Organization (MMO) has incorporated the work previously carried out by the Marine and Fisheries Agency as well parts of the work of the Department of Energy and Climate Change and the Department for Transport. The MMO marked a fundamental shift in how activities in British marine areas are planned, regulated and licensed, with an emphasis on sustainable development. The vision of the MMO is to make a significant contribution to sustainable development in the marine area and to promote the United Kingdom's vision for clean, healthy, safe, productive and biologically diverse oceans and seas. This is an innovative way to bring all maritime activities such as licensing, sustainable development and marine planning together under one roof.

2.4 Chilean fisheries reform

A new law regarding discards was enacted in 2012, including control measures and sanctions for both industrial and artisanal fishing. The implementation process was gradual, starting with an investigation program of minimum two years per fishery, in order to gather technical background that will allow the development of discard reduction plans. These involve regulatory modifications, improvement in gear selectivity, market development, changings in traditional approaches and incentives for the adoptions of changes. The idea behind this new regulation is that given the lack of positive results regarding discards, a new and more realistic approach is needed. The new approach recognises the need for an initial assessment of the problem and then the design and implementation of necessary measures.

In 2013, a new fisheries law was enacted to implement broad reforms. Following the introduction of a TAC system in the early 1990s and an allocation system similar to ITQ by 2001, the new law introduced concepts of precautionary and ecosystem approaches; new definitions and classifications for assessing and measuring the availability of fishery resources; new international sustainability management standards such as Biological Reference Points (BRPs) and Maximum Sustainable Yield; establishment of eleven committees for scientific decisions on the state of fishery resources, BRPs and annual quotas; Conservation Measures for Vulnerable Marine Ecosystems (VMEs) such as seabeds including prohibitions for bottom fishing if it causes damage; obligation for establishing management plans for resources with closed access and recovery plans for overexploited and depleted fisheries; improved rights allocation system same that operates in the same manner as ITQ which may be divisible and last 20 years, can be renewed or terminated depending on owner's cooperation on environment, fishing, labour, etc. In addition, the law established new control elements for larger vessels of more than 12 m length. The first nautical mile from shore is exclusively reserved for small-scale fishing vessels which are less than 12 m length, covering from the northern limit up to the Chiloe Island. GPS is required for small-scale vessels which are more than 15 m length, purse-seine vessels more than 12 m length and all transport vessels. A catch certificate will be required for vessels of more than 12 m length.

Furthermore, the new law brings many rules in line with international norms, such as keeping log books and having specific landing ports for small scale fishers. Additionally, the requirement that landings be certified by a third party was expanded from vessels of more than 18 m to include vessel longer than 12 m as of January 2014. In response to the pressure on wild fisheries resources including jack mackerel, anchovy and sardine as sources for fishmeal and fish oil for salmon aquaculture, the Chilean government is supporting efforts to diversify aquaculture from carnivorous species such as salmon to others like scallops, algae, and mussels.

2.5 Development of aquaculture in the United States

The United States set up a National Strategic Plan for Federal Aquaculture Research covering 2014 to 2019. Meetings of the National Science and Technology Council, the Committee on Science Interagency and the Working Group on Aquaculture were held in June 2014. The second initiative of this working group was a national strategic plan for federal aquaculture research, developed by the National Science & Technology Council's Committee on Science.

The plan includes nine strategic goals with associated outcomes and milestones that identify federal agency and interagency research, science and technology priorities over five years that will support aquaculture development in the United States. The goals are to:

- advance understanding of the interactions of aquaculture and the environment
- employ genetics to increase productivity and protect natural populations
- counter disease in aquatic organisms and improve biosecurity
- improve production efficiency and well-being
- improve nutrition and develop novel feeds
- increase supply of nutritious, safe, high-quality seafood aquatic products
- improve performance of production systems
- create a skilled workforce and enhance technology transfer
- develop and use socioeconomic and business research to advance domestic aquaculture.

As of June 2013, the University of Rhode Island was involved in the first effort in the United States to farm tuna in a land-based aquaculture facility located at the University of Rhode Island's Bay Campus. The university started the first phase of breeding tuna from egg to harvest size. There are many challenges facing tuna aquaculture, including in-tank spawning, feeding the microscopic larvae with live food, then changing to dry formulated feed, and adapting tuna to small spaces.

2.6 Northern Europe

In 2013 the **Danish** Fisheries Minister announced a DKK 10 million (USD 1.7 million) package to help small scale fisheries become more sustainable. Funding is for research carried out at the Danish Technology Institute Aqua (DTU) focussed on the development of gear (particularly Danish seines and gill nets), mapping of coastal fisheries resources, development of seal proof gear and for the sustainable production of mussels.

The Arctic environment ministers met in Jukkasjärvi, Sweden on 5-6 February 2013. The meeting addressed, inter alia, issues related to climate change and ocean acidification, biodiversity and ecosystem services, ecosystem based management and further Arctic cooperation. Increasing ocean temperatures raise the prospects for more fish to be available in this area. Targeted efforts are needed to protect marine biodiversity and Ministers underscored the need to ensure the implementation of agreed biodiversity objectives including the Aichi Biodiversity Targets.

On 12 March 2013 **Norway**, the European Union and the Faroe islands agreed on a 5-year agreement on the management of mackerel in the North East Atlantic. The agreement covers the main part of the stock's distribution area and puts an end to a long-lasting dispute between the parties. Iceland is not a part of the agreement. However the agreement sets aside a considerable share for third parties.

Recreational fishing in **Sweden** involves about a million people, 35% of which are women. Recognising the potential important economic contribution a strategy for the recreational fisheries sector to 2020 was published in 2013. This strategy envisions a growing recreational fishery that is sustainable and available to both urban and rural dwellers, and targets a doubling of fishing tourism by 2020.

The Marine and Coastal Access Act, enacted in 2009, is intended to improve the way the **United Kingdom** uses its marine resources and maximise the benefits it gets from them. Through the Act, United Kingdom Administrations are now in the process of introducing new systems for marine planning and licensing within the policy framework provided by a United Kingdom Marine Policy Statement adopted by all United Kingdom administrations in March 2011. England's first marine plans (for the East inshore and offshore areas) were adopted in April 2014 and, following consultation in 2013, a National Marine Plan for Scotland is expected to be adopted in early 2015. Both Wales and Northern Ireland are working on their national marine plans and are expected to be published for consultation in late 2014 and early 2015 respectively. It is expected that the Marine Policy Statement and Marine Plans will provide a framework for consistent, sustainable, and evidence-based decision making for sustainable marine development.

2.7 Protecting marine ecosystem and combatting IUU fishing in Korea

In 2010, Korea amended the Fisheries Resources Management Act, the basis for establishing Korea's Fisheries Resources Agency (FIRA). FIRA officially began operations in 2011. The goals of this agency are to restore fisheries production to 11 million tonnes per year, to establish 35 000 ha of marine forest area and to be one of the agencies evaluated at the highest level by Korean people by 2030.

Major projects implemented so far include the Marine Forest Enhancement Project designed to enrich marine forest by seaweed planting campaign, the Coastal Marine Ranch project which aims to construct 50 marine-ranch sites by 2020 and research and development on artificial reef development.

In 2013 a national day called "Marine Afforestation Day" was established as 10May, to encourage seaweed planting in areas affected by coral bleaching and to promote the recovery of damaged marine areas.

In October 2014, Korea and China agreed to jointly implement a surveillance programme against IUU fishing in waters shared by the two countries. An initial agreement was reached at the 2013 annual Korea and China fisheries meeting to require Chinese fishing boats to be equipped with automatic identification systems. From 2015 the two countries will perform two to three joint inspection tours during high seasons in fishing.

As of 31 January 2014, the amended *Distant Water Fisheries Act* entered into force to provide increased sanctions including imprisonment of up to three years or a criminal fine of up to three times the potential gain from illegal catches; and extended duration of the suspension on distant water fishing officer's licenses (first offense 30 days, second offense 60 days, third offense revocation). Port inspection was strengthened and installation of VMS on all distant water fishing vessels was made mandatory. To implement stronger monitoring, control and surveillance for Korean-flagged distant water fishing vessels, the Fisheries Monitoring Center was established and started its operation from 20 March 2014.

In January 2015 stronger controls were implemented. This included:

- tighter restrictions on fishing authorization for fishing vessels involved in IUU fishing
- the introduction of an IUU history tracking system
- calling to port, and suspension of operation on vessels under investigation for IUU fishing
- confiscation of illegal catches
- risk-based vessel monitoring and management to prevent IUU fishing.

A new provision was made to allow the Korean government to exercise control over Korean Nationals who have engaged in IUU fishing in Waters outside Korea's jurisdiction taking advantage of flag of convenience. For stronger MCS, mandatory installation of a VMS on carrier vessels and pre-authorization requirements for transhipment were included. Level of sanctions were increased to five years of prison, or a criminal fine of up to five times the wholesale value of the fishery products obtained from relevant violations, based on the average wholesale prices of the products for the preceding three years; or a criminal fine at least KRW 500 million (USD 474 833) up to KRW 1 billion (USD 949 667), whichever is higher.

As of February 2015, United States removed Korea from the list of countries potentially engaged in IUU fishing. NOAA included Korea on the list in 2013, citing inadequate restrictions on Korean vessels' illegal fishing in the Antarctic.

2.8 Japan's effort to promote fish consumption

Contrary to global trends, Japan has been experiencing declining fish consumption. In 2009, Japan's human consumption of fisheries products per capita (54 kg) was ranked as the third in the world, following Portugal (61kg), and Korea (56 kg). According to a survey on seafood consumption trends in 2006, the reasons for not eating fish were as follows: (1) family members' preference; (2) higher price compared to meat; (3) little knowledge; (4) individual preference; and (5) difficulty in cooking fish.

To expand fish consumption, the "Delight of a Fish-Rich Country" project as a public-private collaboration effort was launched in 2012. Under this program, a system was established to seek handy and tasty processed fishery products and seasoning from the public and select "Fast Fish" products from among them.

2.9 European Maritime Fisheries Fund

The EMFF (European Maritime and Fisheries Fund) was approved by the EU Parliament and Council in May 2014. It is a EUR 6.7 billion financial support instrument for the programming period of 2014 until 2020. The EMFF focuses on providing financial support for the implementation of the Common Fisheries Policy for the conservation of marine biological resources, for the management of fisheries and aquaculture, for the European Union's integrated Maritime Policy as well as data collection, control, governance and market policy in the fisheries sector. Funding provided by the EMFF will enable fishing and aquaculture activities to contribute to the creation of environmental conditions capable of being sustained long-term which are necessary for economic and social development.

References

OECD/FAO (2015), *Fishing for Development*, The Development Dimension, FAO Publications, Rome.
dx.doi.org/10.1787/9789264232778-en.

OECD (2011), *Towards Green Growth*, OECD Green Growth Studies, OECD Publishing, Paris. dx.doi.org/10.1787/9789264111318-en.

OECD (2006), *Fishing for Coherence: Fisheries and Development Policies*, The Development Dimension, OECD Publishing, Paris.
dx.doi.org/10.1787/9789264023956-en.

Chapter 3

International developments in fisheries policy

This chapter describes recent developments in some important global fisheries producers and looks at international developments in fisheries policy. There is a continued emphasis on improving the sustainability of capture fisheries and on promoting the continued rapid growth of the aquaculture sector in many countries. Trade negotiations continue to expand, with many agreements being signed or completed in the past two years. Regional Fisheries Management Organisations (RFMOs) continued working towards improved effectiveness, in particular for curbing illegal, unreported or unregulated fishing.

3.1 Recent developments in key partner economies

Brazil

In May 2012, two new sanitary measures were introduced in Brazil, namely, the National Hygiene and Sanitary Control of Bivalve Molluscs (PNCMB) and the National Network of Laboratories of the Ministry of Fisheries and Aquaculture (Renaqua). The PNCMB is in charge of the monitoring production for human consumption. The Ministry of Fisheries and Aquaculture (MPA) is responsible for observing and surveying areas of cultivation and origin of shellfish like oysters, mussels, cockles and viera. The Ministry of Agriculture, Livestock and Supply (MAPA) defines criteria for hygiene and sanitary industrial processing and inspects establishments linked to the Federal Inspection Service (SIF).

The National Network of Laboratories is initially composed of four laboratories, i.e. the Federal University of Minas Gerais (UFMG), State University of Maranhão (UEMA), the Federal Institute Santa Catarina (IFSC) and the Integrated Agricultural Development Company of Santa Catarina (Cidasc). This network will undertake official analyses and diagnoses. The network will also be responsible for the development of new technologies for examination of diseases, residues and contaminants.

To develop technologies in fisheries and aquaculture, the Ministry of Fisheries and Aquaculture (MPA) and the Ministry of Science, Technology, and Innovation (MCTI) agreed to form an inter-ministerial technical committee in January 2013. This committee aims to have a strategic role in research covering a wide range of issues including breeding, cultivation and management, equipment, feed, assessment of fish stocks, bio-economy, fish processing, storage, transportation and marketing. This co-operation between the two ministries also includes sharing resources of MCTI with funding from the MPA for research projects in the above areas. Brazil aims to develop itself as one of the world's largest producers of fish similar to what has taken place for meat and grains.

The Brazilian Government has been implementing "The harvest plan for fisheries, and aquaculture 2012/2013/2014" with an investment of BRL 9.8 billion (USD 4.1 billion) for expanding aquaculture, modernising and strengthening the fishing industry and fisheries trade. The goal of this plan is to attain a production level of 2 million tonnes of fish by the year 2014.

Brazil is also focusing on the sustainable utilisation of fisheries waste. Two plants producing biodiesel by using tilapia waste were opened in Jaguaribara and Morada Nova in March 2013. The National Department of Works Against Drought (DNOCS) announced this new facilities will produce around 8 000 litres of biodiesel per day. These plants are expected to provide added value and sustainability to the tilapia fisheries producers.

People's Republic of China

Following the 12th Five-year (2011-2015) Plan for Chinese Fishery released by Ministry of Agriculture in 2011, a dynamic fishing vessel control system is being built by establishing a national fishing vessel data-base. Different government agencies are coordinating inter alia fishing vessel inspection and registration, fishing licence issuance, establishment and checks for restrictions on fishing gears. Cross-agency communication and cooperation is encouraged to improve integrated fisheries management. The Plan includes optimisation of location with regard to ocean industries, restructuring of traditional ocean industries and improvements that include the relocation, standardisation and greening of aquaculture farms. The People's Republic of China (hereafter "China") is also developing deep sea cage aquaculture and recirculating aquaculture systems.

India

In India, aquaculture is the main driver for growth in fish production. Aquaculture grew at an average of 5% over the last decade compared to 4.3% annually overall for fisheries. It is forecasted to continue to grow at around 2% per year. Fish is the largest source of protein for Indian consumers with fish consumption at 5.9 kg per capita in 2013. It is expected to grow 1% every year to reach 6.8 kg by 2023. India's aquaculture overtook capture fisheries in terms of production in 2013.

The National Fisheries Development Board (NFDB) was established in August 2014 in order to have a more integrated fisheries governance system. The NFDB will promote the fisheries sector and co-ordinate activities related to fisheries undertaken by different ministries or departments in the central government and state or union territory governments. The NFDB will also carry out research with a view to improving production, processing, storage, transport and marketing of fisheries and aquaculture products. It will also provide standards for sustainable management and conservation of natural aquatic resources including for fish stocks. Among other activities this will be coordinated with the private sector, banks, financial institutions, and others for promoting development partnerships in the fisheries sector.

The increasing importance of the aquaculture industry is bringing food safety to the fore. The first Aquaculture Lab was accredited by the Indian National Authority in 2014. With the decision by the Accreditation Board for Testing & Calibration Laboratories (NABL) as per ISO/IEC 17025: 2005, the Central Aquaculture Pathology Laboratory of the Rajiv Gandhi Centre for Aquaculture (RGCA), under Marine Products Export Development Authority (MPEDA) will perform as the sole accredited aquaculture lab.

Indonesia

The fishing industry plays a crucial role in the national economy; 6.5 million Indonesian people depend on marine and inland water fisheries and aquaculture activities.

The Indonesian and Dutch governments are collaborating to improve food security. In November 2013, the Indonesian Minister of Marine Affairs and Fisheries and the Dutch Minister for Agriculture signed an agreement to increase domestic fisheries production and reduce harmful fishing practices. Both countries committed to contribute a total of EUR 9 million (USD 12 million). Based on this agreement, the Project for Fisheries and Aquaculture for food Security in Indonesia will be implemented over the three year period 2014 to 2016. The project will be managed by the Center for Development and Innovation (CDI) of Wageningen University with a budget of EUR 4.5 million (USD 6.21 million). Currently, around a third of the catch in Indonesia is lost due to weaknesses in the treatment, storage, distribution, and marketing of fish.

Indonesia began preparing a Master Plan for Aquaculture Development in June 2014. The government invited the WorldFish Centre to work jointly on a master plan for national aquaculture by 2020. The plan will be drafted through research projects over the following 18 months. Tasks of this project include developing scenarios of supply and demand for fishery products, and building an understanding of the opportunities and challenges to stimulate sustainable aquaculture in Indonesia.

South Africa

Abalone farms have created the majority of direct new jobs in the aquaculture sector in recent years. Abalone aquaculture itself accounts for more than half of total aquaculture production and more than 95% of the total value of production. Their high value has led to IUU fishing activities resulting in abalone stocks being endangered. To tackle overfishing of abalone and other species as well as ease the pressure on global fish resources, the South African government has been promoting the aquaculture industry.

To develop the aquaculture industry the Aquaculture Development and Enhancement Programme (ADEP) was established in 2013 by the Department of Agriculture, Fisheries and Forestry in co-operation with the Department of Trade and Industry. The goal of the ADEP is to encourage investments in the aquaculture sector to increase sustainable production and the creation of jobs and skills development. A funding incentive system was put in place to attract new farmers into the aquaculture business. According to the Department of Trade and Industry, ADEP would provide a significant grant to a maximum of ZAR 40 million (USD 3.6 million) for new and expansion projects.

3.2 International developments

Trade agreements

Many OECD countries actively participated in the negotiation and conclusion of Free Trade Agreements (FTAs) in 2012 and 2013, including the European Union, Japan, Korea and the United States.

The EU-Korea FTA entered into force on July 2011. In 2012, the European Union launched FTA discussions with Japan, and Viet Nam. In 2013, talks with the United States and Thailand were initiated. EU negotiations of FTAs finalised with Peru in March 2013 and with Colombia in August 2013. In July 2014 negotiations were concluded for the accession of Ecuador to the Trade Agreement with Colombia and Peru. In August 2014, Canada and the European Union concluded the negotiations of the Canada-EU Trade Agreement. The European Union has also been negotiating Economic Partnership Agreements (EPAs) with Africa, Caribbean, Pacific (ACP) countries. This group is composed of seven regions: five in Africa, one in the Caribbean and one in the Pacific. Most recently, a deal was reached in July 2014 with 16 West African states and in October 2014 with the East African community. As of the end of October 2014, the European Union is also negotiating with Mercosur on a possible future FTA.

The European Union and the United States are continuing negotiations under the goal of finalising the Trans-Atlantic Free Trade Agreement by the end of 2016. The European Union expects that liberalising trade across the Atlantic could add an additional USD 160 billion a year to GDP in Europe. This FTA once established will be the world's largest free-trade area.

As of July 2013, Japan concluded FTAs or Economic Partnership Agreements (EPAs) with 13 economic entities including Singapore, Mexico, Malaysia, Chile, Thailand, Indonesia, Brunei, ASEAN, Philippines, Switzerland, Viet Nam, India, and Peru. At that time, Japan was negotiating with Australia, Mongolia, Canada, Colombia, China, European Union, RCEP, TPP, GCC, and Korea. In April 2014, Japan agreed with Australia to sign a free trade agreement, following seven years of negotiation.

Korea is also actively involved in FTA negotiations with various partners. The Korea-Chile FTA has been in force since 2004. Korea also has FTAs with Singapore, EFTA (The European Free Trade Association; Iceland, Liechtenstein, Norway, Switzerland), European Union, and United States. In 2012, Korea signed a FTA with Colombia.

In the early 2014, Korea signed FTAs with Australia and Canada. As of July 2014, Korea is pursuing negotiations with Viet Nam, Indonesia, and China. The FTA with China in particular is expected to have a large impact on the Korean fisheries industry.

RFMO developments

Many RFMOs are working on improved procedures for combating IUU fisheries. The ICCAT (International Commission for the Conservation of Atlantic Tunas) has been working since 2011 to transfer catch certification from paper to electronic format to improve transparency of capture and trade of Bluefin Tuna. For Atlantic Tuna in the ICCAT region, the Electronic Bluefin Tuna Catch Document (E-BCD) became fully operational in 2014. Furthermore, ICCAT continues to work toward the inclusion of observers by 2015. It also seeks to expand the number of species covered, including sharks for example.

The Indian Ocean Tuna Commission (IOTC) adopted new conservation and management measures in 2013. The IOTC has been checking compliance of member states with the conservation and management measures adopted by the commission. IOTC is targeting in particular abuse of fish aggregating devices (FADs). IOTC has been collecting member states' management plans for FADs used by tuna purse seine fishing vessels and bait boat vessels since 2013. It is also managing the IOTC IUU fishing vessel list in the convention area.

The Southern Indian Ocean Fisheries Agreement (SIOFA) was established in June 2012. SIOFA manages fisheries resources other than tuna and tuna-like species in the High Seas of Southern Indian Ocean. Founding member states are Australia, Cook Islands, the European Union, France on behalf of its overseas territories, Mauritius and the Seychelles.

Discussion on the establishment of a North Pacific Fisheries Commission has been on-going since 2006. The Convention text on the Conservation and Management of High Seas Fisheries Resources in the North Pacific Ocean was adopted in February 2012. The Commission covers all species except tuna and tuna like species, salmon, and pollock in the North Pacific Area. Species of particular interest include Pacific saury, Neon flying squid, and splendid alfonsin. Members agreed to have a Secretariat in Tokyo. Participants are Canada, China, Japan, Korea, Russian Federation, United States, and Chinese Taipei.

On 24 August 2012, The Convention on Conservation and Management of the High Seas Fishery Resources of the South Pacific Ocean entered into force. Members of the South Pacific Regional Fisheries Management Organization (SPRFMO) are Australia, Belize, Chile, China, Cook Islands, Cuba, European Union, Denmark in respect of the Faroe Islands, Korea, New Zealand, Russian Federation, Chinese Taipei and Vanuatu. This organisation is meant to fill the gap that existed in the international conservation and management of non-highly migratory fisheries and protection of biodiversity in the marine environment in high seas areas of the South Pacific Ocean The main species covered by this RFMO are pelagic fisheries for Jack mackerel and bottom fisheries for species like Orange roughy. The first SPRFMO meeting was held from 28 January to 1 February 2013 in Auckland, New Zealand.

As a global programme co-ordination unit, the FAO is implementing the Areas Beyond National Jurisdiction (ABNJ) programme with the Global Environment Facility (GEF), UNEP, World Bank (hereafter WB) as well as RFMOs, industry and NGOs. This project aims to protect *vulnerable* and important ecosystems and species living in the areas beyond national jurisdiction including the high seas and areas of the seabed beyond the continental shelf of coastal states. The area accounts for 40% of surface of planet, and 64% of the surface of the oceans, and nearly 95% of oceans' volume. There are four sub-projects namely: (1) Sustainable Management of Tuna Fisheries and Biodiversity Conservation in ABNJ; (2) Sustainable Fisheries Management and Biodiversity Conservation of Deep-sea Living Resources and Ecosystems in ABNJ; (3) Ocean Partnerships for Sustainable Fisheries and Biodiversity Conservation-Models for Innovation and Reform; and (4) Strengthening Global Capacity to Effectively Manage ABNJ.

Trans-Atlantic Ocean Research Alliance

Addressing research needs for the Atlantic Ocean, a systematic co-operative network was formed between the European Union, Canada and the United States in 2013. The Galway Statement on Atlantic Ocean Cooperation was signed and a research alliance was launched in May 2013 at the Marine Institute, Galway, Ireland. This historical agreement focused on aligning ocean observations and research among the three participants. Other potential areas for co-operation include: (1) sharing of data, such as on temperature, salinity and acidity; (2) interoperability and coordination of observing infrastructures, such as measurement buoys and research vessels; (3) sustainable management of ocean resources; (4) seabed and benthic habitat mapping; (5) promoting researcher mobility; and (6) identifying and recommending future research priorities.

In 2013, Marine and Arctic working groups were established after the launch of the Trans-Atlantic Ocean Research Alliance. Research on aquaculture is one of the main issues in the Alliance. For example, the objective of Canada-EU marine group is to progress work on selected marine issues under the EU-United States Science and Technology Agreement including Ocean Stressors, Aquaculture, Observing Systems, Marine Microbial Ecology, and Ocean Literacy.

FAO Committee for Fisheries meeting

The FAO Committee for Fisheries held its 31st biannual session in Rome on the 9-13 June 2014. The major outcome of the meeting was the endorsement of the Voluntary Guidelines for Securing Sustainable Small-scale Fisheries in the Context of Food Security and Poverty Eradication (SSF Guidelines). The SSF Guidelines are intended to "promote the implementation of national policies that will help small scale fishers thrive, and play an even greater role in ensuring food security, promoting good nutrition, and eradicating poverty."

The SSF Guidelines, which have been under development since 2011, emphasise human rights and dignity, highlight the need for gender equality and encourage countries to ensure small-scale fishers are involved in decision-making that affect their livelihoods. The SSF Guidelines are voluntary and their impact will come from helping small-scale fishers understand their rights and by influencing policy and the policy-making process, in particular for developing countries.

In conjunction with the 31st session, the State of World Fisheries and Aquaculture (SOFIA) 2014 publication was released. This flagship biannual publication provides a comprehensive global overview of the state of the world's capture fisheries and aquaculture via a detailed reporting of data.

The Committee also endorsed the Voluntary Guidelines for Flag State Performance (VGFSP). These Guidelines identify actions that countries can take to ensure that vessels registered under their flags do not conduct IUU fishing. The Voluntary Guidelines aim to eliminate the practice of "flag hopping", where vessels repeatedly re-register with new flag States to escape detection of their illegal activities. It also promotes greater cooperation and information exchange between countries, so that flag states are in a position to refuse to register vessels that have previously been reported for IUU fishing, or that are already registered with another flag state. The guidelines also provide recommendations on how to encourage compliance, as well as on how to assist developing countries to fulfil their flag state responsibilities.

Part II

OECD AND NON-OECD ECONOMY PROFILES

Part II of the Review provides summary descriptions of recent developments in OECD countries and partner economies, and contains detailed individual country profiles in the electronic version.

Chapter 4

OECD and non-OECD economy snapshots

This chapter provides summary descriptions of recent developments in OECD countries and partner economies.

ARGENTINA

Summary and recent developments

- In 2013, exports of fish products reached an all-time high, recovering strongly from the decline observed in 2012. As usual, the performance of the shrimp and squid fisheries strongly determines the overall value of exports. Low export prices (except for shrimp) and high costs have put pressure on the fishing sector. The global financial crisis is still affecting traditional export markets and new markets are demanding products at lower prices.
- Governance Strengthening for the Management and Protection of Coastal- Marine Biodiversity in key ecological areas and the implementation of the Ecosystem Approach to Fisheries (EAF) is a new project designed to strengthen management capacities and protection of coastal-marine biodiversity in key ecological areas by creating new Marine Protected Areas (MPAs) and implementing the Ecosystem Approach to Fisheries (EAF) including in existing management plans of MCPAs.
- The Undersecretariat of Fisheries and Aquaculture and the National Agri-food Health and Quality Service (SENASA) have begun the development of a traceability system for fishery products.
- The Agricultural Strategic Plan (PEA2020) is being jointly prepared by government agencies, private sector, NGOs, universities and research centres. Aquaculture has been included in the strategic plan since 2011 and the marine fishing sector was integrated in 2013.
- The Pampa Azul Initiative was presented in 2014. This project of scientific research in the Argentine sea includes activities related to exploration and conservation, technological innovation for the productive sectors related to the sea and scientific communication addressed to the public in general. It is coordinated by a number of government ministries. The objectives are to deepen scientific knowledge as a basis for the conservation and management of natural resources, to promote new innovations for the sustainable use of natural resources and to develop marine industries.

Capture fisheries and aquaculture production

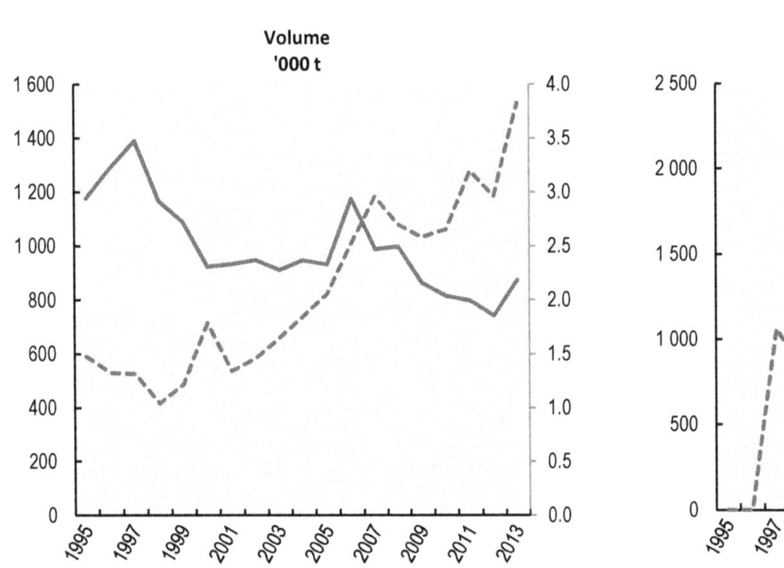

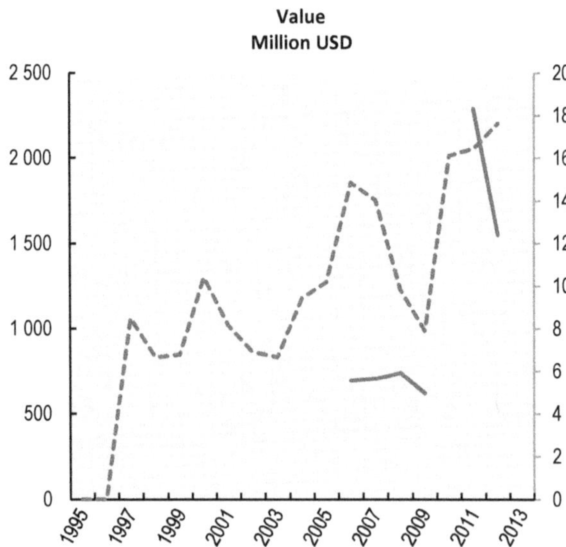

Source: FAO FishStat Database.

Key characteristics of Argentinian fisheries

- Total landings from 2011 to 2012 showed a slightly decline from 733 000 tonnes to 692 072 tonnes and a considerable growth in 2013 to 822 067 tonnes, leaded by an extraordinary squid harvest (Panel A).
- About 90% of marine fisheries products, in terms of landings, are exported and there is a large positive trade balance. Exports reached a new high in value terms in 2013 (Panel B).
- While general services dominate the GFT, there is a direct payment programme where a government subsidy to worker's wages is available to help prevent job losses (Panel C).
- In 2013 the maritime fishing fleet was composed by 902 vessels, 184 of them have only provincial permits, 235 have both, national and provincial, the others have only national permits. Employment in fishing sector dropped 2% between 2012 and 2013, with job losses concentrated in the processing sector. In 2013 there were 13 884 workers employed in marine fishing, 7 575 in processing and 7 000 in inland fisheries (Panel D).

Key fisheries indicators

Panel A. Key landings in value, 2013

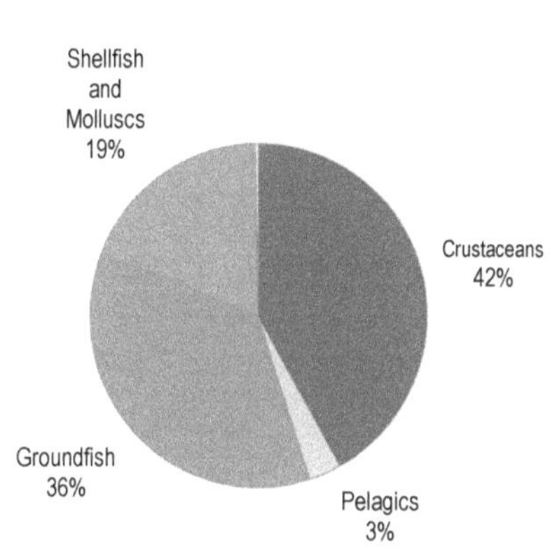

Panel B. Trade evolution

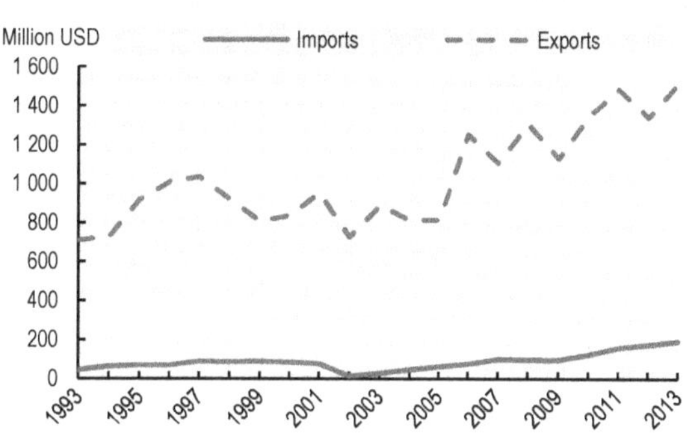

Panel C. Evolution of government financial transfers for marine capture fisheries

USD	Average 2005-10*	2012	% change
Direct payments	..	4 707 867	..
Cost reducing	..	..	..
General services	..	28 722 791	..

* Or closest available years.

Panel D. Capacity

	Average 2005-10*	2013	% change
Number of fishers	16 890	13 884	-17.8
Number of fish farmers	290	1 238	327
Total number of vessels	1 044	902	-14
Total tonnage of the fleet	193 281	179 806	-7

* Or closest available years.

AUSTRALIA

Summary of recent developments

- In 2011-12 the gross value of Australian fisheries production increased 3 per cent to AUD 2.3 billion (USD 2.3 billion) mainly thanks to increase of aquaculture production by AUD 100 million to AUD 1.1 billion which account for 46% of the total Australian fisheries production. In volume terms, Australian fisheries production increased slightly, by 476 tonnes to 237 540 tonnes. The volume of aquaculture production increased by 10% from 75 188 to 84 605 tonnes, accounting for 36% of total Australian fisheries production while that of capture fisheries declined by 4% to 157 505 tonnes.
- Sustainable development is supported by the production of annual Fishery Status Reports which provides an independent evaluation of the biological, economic and environmental status of fish stocks. Stocks improved overall in 2013, compared with previous years. It is the first year since 2006 that no stocks managed solely by the Australian Government have been classified as subject to overfishing.
- In 2014, Australia released its National Aquaculture Statement recognising the contribution that aquaculture makes to the Australian economy and regional development. The statement is the first step in fulfilling the Australian Government's commitment to work with industry to develop a national aquaculture strategy .
- In 2013, Australia hosted the first meeting of parties to the Southern Indian Ocean Fisheries Agreement (SIOFA). Parties to the agreement focused on developing foundation documents, including rules of procedure and financial regulations to help ensure the long-term conservation and sustainable use of non-highly migratory fish stocks in the high seas of the southern Indian Ocean.

Capture fisheries and aquaculture production

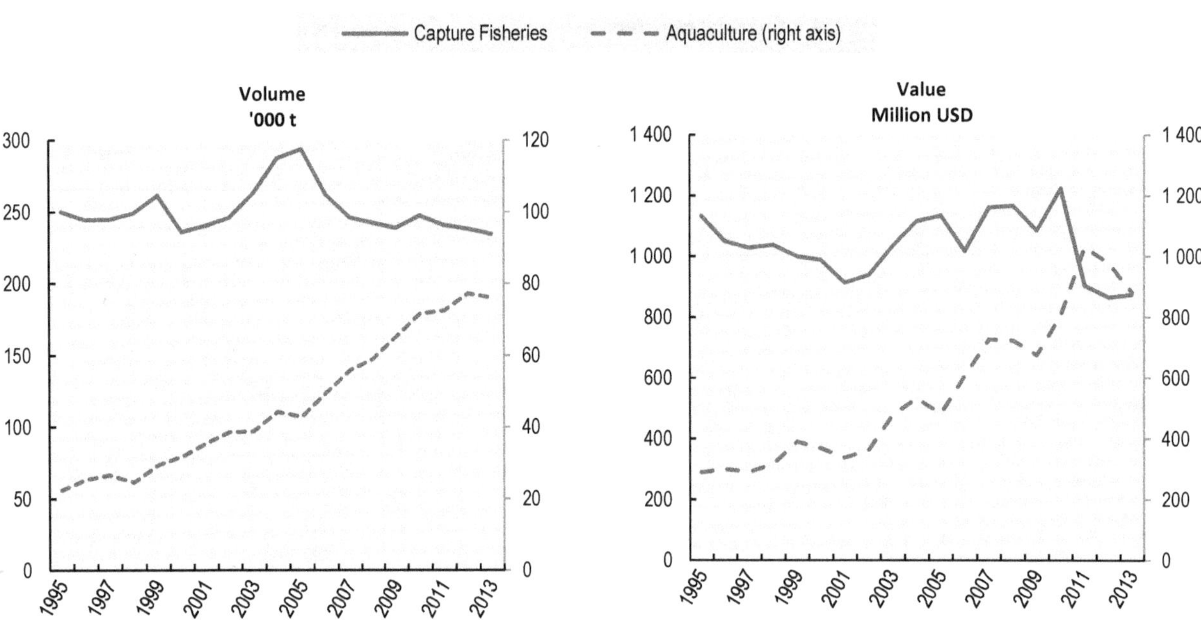

Source: FAO FishStat Database.

Key characteristics of Australian fisheries and aquaculture

- The gross value of Australia's capture fisheries production decreased in recent years from both state and Commonwealth fisheries. Crustaceans continue to be the most important species landed in 2012 in terms of value, followed by shellfish and molluscs and tuna. (Panel A)
- While imports continue to be a significant feature of domestic fish product consumption, Australia's exports of high value fish products such as tuna, lobster and abalone continue to grow. el B)
- A total of AUD 44.17 million was transferred to the Australia's fisheries sector in 2011-2012 (AUD 13.82 million net of cost recovery charges), which is slight decrease compared to that of 2010 (AUD 45 million). (Panel C)
- The number of fishing vessels and tonnage remained stable between 2010 and 2013 after a significant decline between 2007 and 2009. The number of people employed in the fisheries sector decreased 19% between 2011-12 and 2012-13. (Panel D)

Key fisheries indicators

Panel A. Key landings in value, 2013

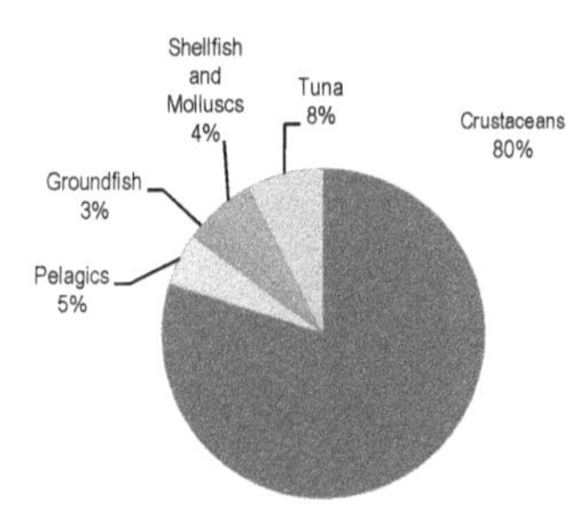

Panel B. Trade evolution

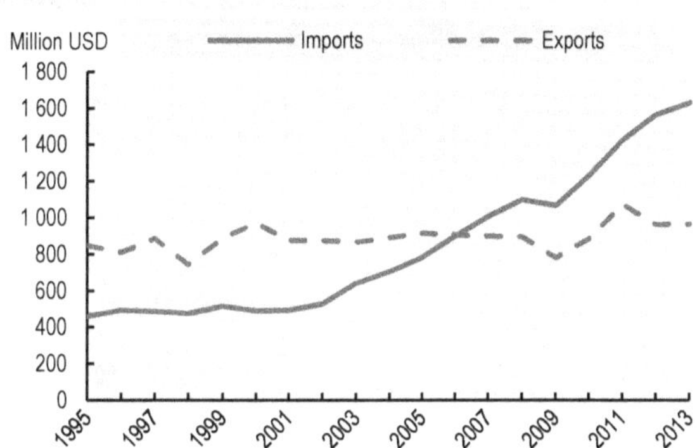

Panel C. Evolution of government financial transfers for marine capture fisheries

USD	Average 2005-10*	2013	% change
Direct payments	..	..	..
Cost reducing	..	..	..
General services	46 276 852	43 078 000	-7

* Or closest available years.

Data for 2013 presents information concerning the fiscal year 2012-13.

Panel D. Capacity

	Average 2005-10*	2013	% change
Number of fishers	7 013	5 050	-28
Number of fish farmers	3 993	3 558	-11
Total number of vessels	453	306	-32
Total tonnage of the fleet	49 443	35 713	-28

* Or closest available years.

BELGIUM

Summary and recent developments

- In 2013 the Belgian fleet consisted of 80 vessels with a total capacity of 46.525 kW (sq) and 14.645 GT. The situation of the Belgian sea fisheries fleet changed significantly during the year 2009 due to a vessel scrapping program organised by the government.
- The total catch of fishery products by Belgian vessels in 2013 increased by 4% to 22 793 tonnes compared to 2012. Seventy-two per cent of this amount was landed in Belgian ports and the remaining abroad. The average price decreased in 2013 by 8% to EUR 3.21/kg. The total value of the catches in both Belgian and foreign ports amounted to EUR 73.1 million (-4%) in 2013.
- The Belgian fleet consists almost exclusively of demersal trawlers. In both 2012 and 2013, more than 90% of the catches were demersal species. Amongst them, sole is the most important species by value. In 2013, landings of sole represented 35% of the overall value of all landings by Belgian vessels.
- After the serious drop in fuel price in 2009 with a decrease of 35% compared to 2008, the fuel prices increased again to average 0.74 EUR/l in 2012 and decreased slightly with 0.04 EUR/l to 0.70 EUR/l in 2013.

Capture fisheries and aquaculture production

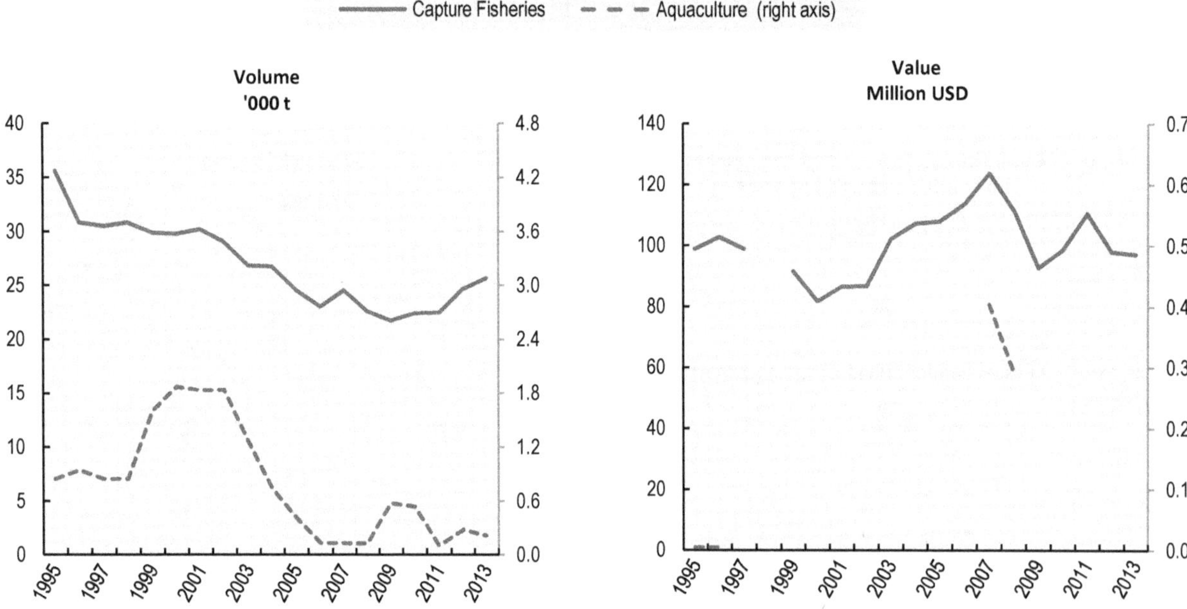

Source: FAO FishStat Database.

Key characteristics of Belgium fisheries

- In 2013, 22 800 tonnes of fish were taken by the Belgian fleet with flatfish being the economically most important species accounting for 68% of the value of landings (Panel A).
- In 2013 imports of fish and fish products amounted to EUR 1 644 million. A substantial part of the national landings, and of fish imports, is exported. This mainly concerns sole, cod, whiting and plaice but also foreign fisheries products like salmon, tuna and pangasius. The major export markets are the Netherlands, France, Denmark, Germany, United Kingdom and Spain. Exports amounted to EUR 834 million (Panel B).
- A total of EUR 4,35 million was spend on the fisheries and aquaculture sectors in 2013 of which the main part of (EUR 3,4 million or 78%) was for general management services. Grants for vessel modernisation and equipment amounted to EUR 850 000 (Panel C).
- The number of vessels in the Belgian fleet stood at 80 in 2013, three less than in the previous year and down from 100 vessels in 2008. The direct employment in the fisheries sector is approximately 2 500 persons with around 400 to 500 persons employed on vessels. Approximately 1 400 persons are employed in the fisheries processing sector (Panel D).

Key fisheries indicators

Panel A. Key landings in value, 2013

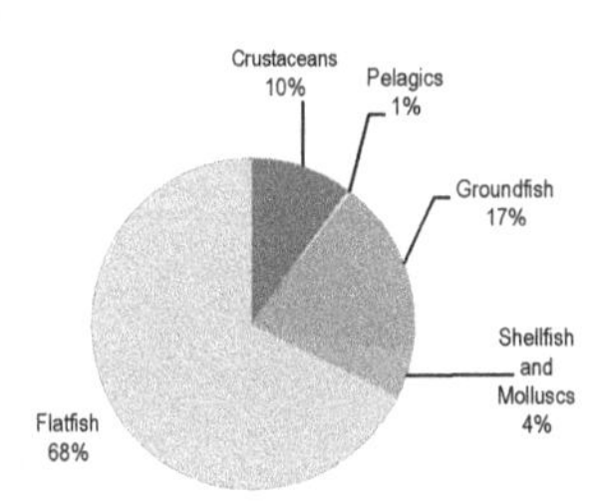

Panel B. Trade evolution

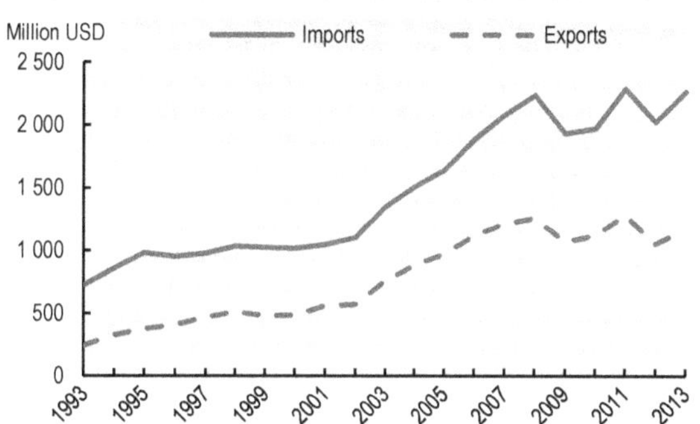

Panel C. Evolution of government financial transfers for marine capture fisheries

USD	Average 2005-10*	2013	% change
Direct payments	7 162 567	1 129 749	-84
Cost reducing	..	..	..
General services	..	4 548 994	..

* Or closest available years.

Panel D. Capacity

	Average 2005-10*	2013	% change
Number of fishers	742	396	-47
Number of fish farmers	143	..	..
Total number of vessels	107	80	-25
Total tonnage of the fleet	19 953	14 645	-27

* Or closest available years.

CANADA

Summary and recent developments

- The *Fisheries Act*, Canada's primary statute for the conservation of fish and fish habitat and the management of fisheries resources, was amended in 2012. The amendments focus the Act's regulatory regime for the protection of fish and fish habitat on managing threats to the sustainability and ongoing productivity of Canada's commercial, recreational and Aboriginal fisheries. They also provide enhanced compliance and protection tools; provide for clarity, certainty and consistency of regulatory requirements through the use of standards and regulations; and enable partnerships with agencies and organisations best placed to provide fisheries protection services.
- In 2014, Canada completed annual surveys of 155 major fish stocks using its Fishery Checklist questionnaire for 2013. One of the questions in the Fishery Checklist asks for the status of the stocks. Seventy-four stocks were in the healthy zone, above their upper stock reference point; 41 were in the cautious zone between their biological limit reference point and their upper stock reference point; 16 stocks were in the critical zone below their biological limit reference point; and the status of 24 stocks was unknown. Since 2011, knowledge about the status of the major stocks has improved, with 11 fewer stocks in the unknown category.
- For 2014-15, Fisheries and Oceans Canada has identified three organisational priorities relevant to fisheries policy, including the improvement of fisheries management through industry-led, incremental fisheries management reforms and improved access to export markets for Canadian products, advancing policy and programme changes to ensure the long-term sustainability of Canada's aquatic ecosystems, and advancing management and operational excellence to modernise the design and delivery of programmes and services.

Capture fisheries and aquaculture production

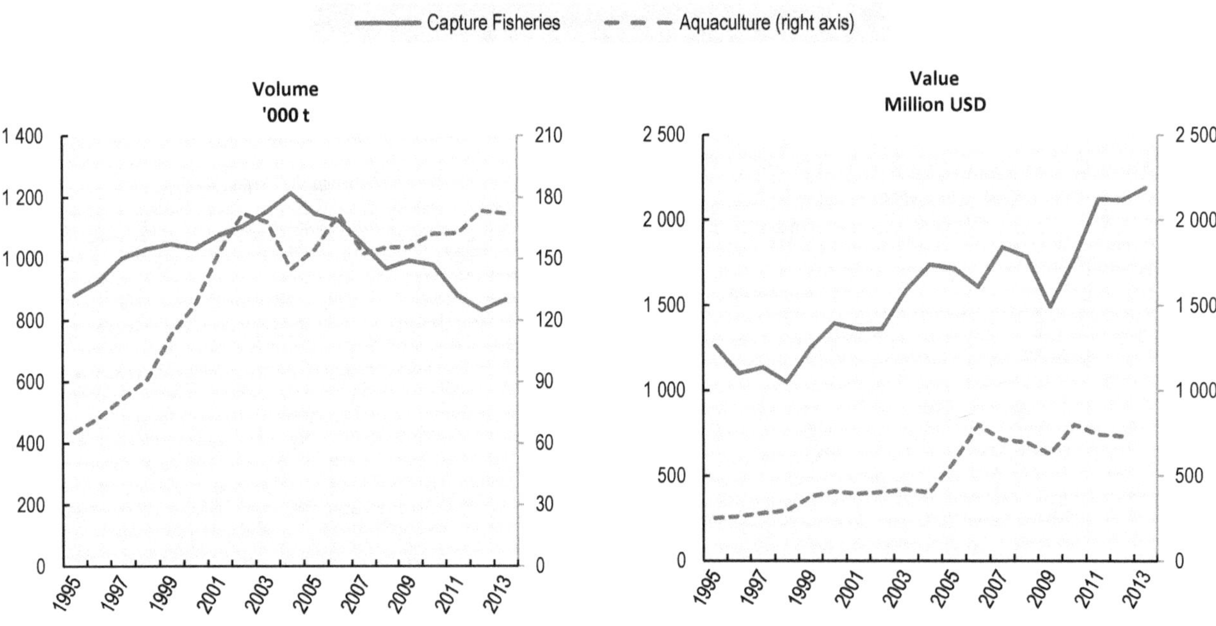

Source: FAO FishStat Database.

Key characteristics of Canadian fisheries

- More than 80% of Canadian landings by value are from the Atlantic fishery. Crustaceans are the most important product, primarily lobster, shrimp and crab. From 2011 to 2013, landed quantities decreased by about 2% but values increased by around 7%. (Panel A)
- Canada is a net exporter of fish products. Its most important export by value is crustaceans and the United States is the largest export market. The value of exports and imports rose steadily between 2011 and 2013, whereas the quantity of both exports and imports fell during the same period (Panel B).
- In 2013, government financial transfers (GFTs) to the fisheries sector were estimated at CAD 718.6 million, representing a return to levels consistent with the 2005 to 2008 period. GFTs represented 22% of the total landed value of Canadian fisheries (including production from aquaculture). The provision of infrastructure and unemployment insurance accounted for 51.5% of the total support (Panel C).
- Approximately 43 000 people were employed as commercial fish harvesters in 2012, with a further 33 000 employed in seafood product preparation and packaging. In 2012, there were 18 740 registered vessels in Canada (Panel D).

Key fisheries indicators

Panel A. Key species by value in 2012

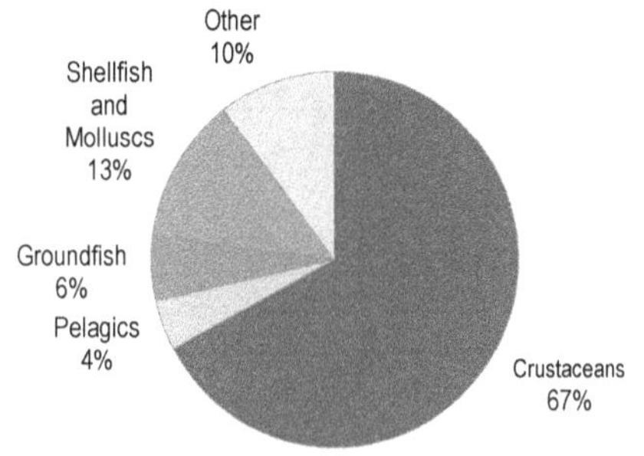

Panel B. Trade evolution

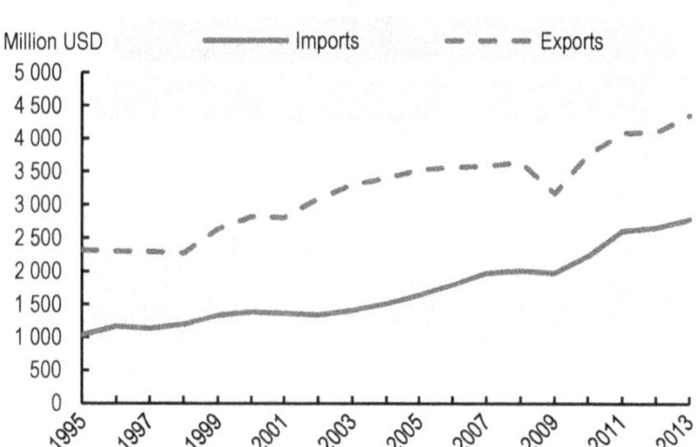

Panel C. Evolution of government financial transfers

USD	Average 2005-10*	2012	% change
Direct payments	266 127 441	..	..
Cost reducing	10 448 505	..	..
General services	420 637 090	..	..

* Or closest available years.

Panel D. Capacity

	Average 2005-10*	2012	% change
Number of fishers	51 147	43 250	-15.4
Number of fish farmers	3 763	3 235	-14
Total number of vessels	21 409	18 740	-12
Total tonnage of the fleet	..	..	..

* Or closest available years.

CHILE

Summary and recent developments

- The General Law of Fisheries and Aquaculture was reformed in 2013. Its main objective is the sustainability of the fisheries, with express reference to the ecosystem approach and the precautionary principle. The Law includes definitions of Biological Reference Points and Maximum Sustainable Yield, a decision making process based on scientific advice, a mandate to elaborate management plans for closed access fisheries and rebuilding plans for overexploited and depleted fisheries.
- A new rights allocation system has been put in place, called Transferable Quota Licenses, similar to ITQ systems.
- In order to prevent harmful, environmental, and sanitary conditions in aquaculture, land planning and relocating concession groups (called neighbourhoods) has been undertaken. New environmental and sanitary measures include the establishment of neighbourhoods to contain diseases by restricting transport between groups in case of emergency. A regulation on densities that limits the number of fish in a farming centre as a consequence of poor environmental and sanitary performance during the previous productive period has also been put in place.
- The volume of aquaculture exports surpassed that of capture fisheries for the first time in 2012.

Capture fisheries and aquaculture production

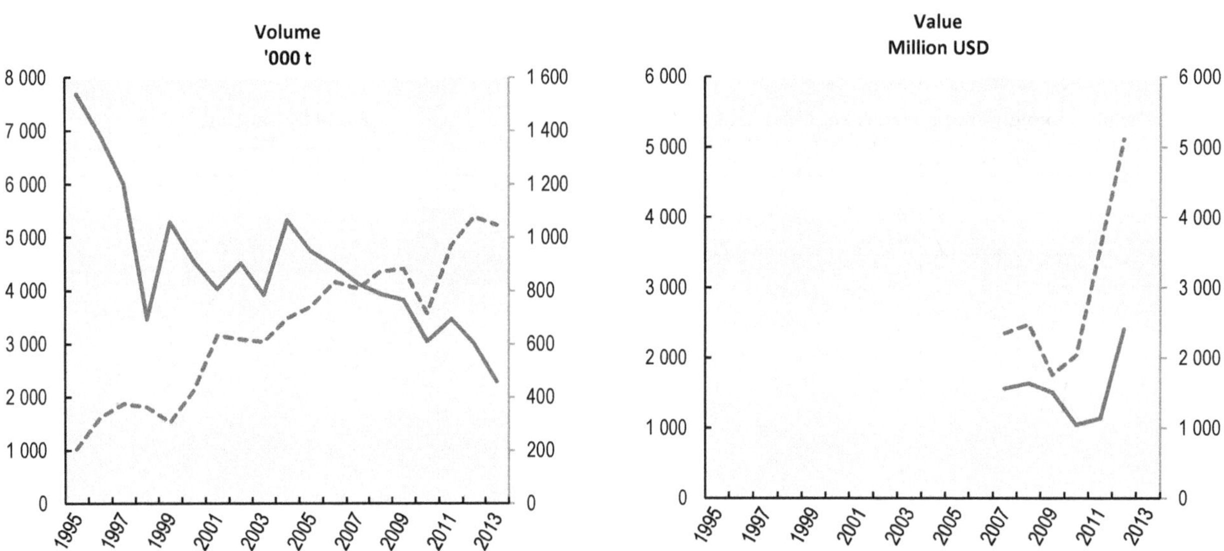

Source: FAO FishStat Database.

Key characteristics of Chile fisheries

- Total landings have trended downward, from 3.8 million tonnes in 2009 to 2.3 million tonnes in 2013. Anchovy make up 40% of landings for small-scale fishers and 60% of landings for industrial fishers (Panel A).
- Export of aquaculture products was 656 000 tonnes in 2012, while for capture fisheries it was 603 000 tonnes (Panel B).
- There are two main public funds that provide support to the fisheries sector in Chile: The Fisheries Management Fund (Fondo de Administración Pesquera, FAP) and the Fund for Development of Artisanal Fisheries (Fondo de Fomento a la Pesca Artesanal, FFPA). Much support is targeted for the development of the small-scale fishing sector (Panel C).
- The industrial fleet and the aquaculture sector provided 2 150 and 11 000 jobs respectively in 2012. Processing plants generated around 34 000 jobs. In terms of small-scale fisheries, 91 000 fishermen are currently registered (Panel D).

Key fisheries indicators

Panel A. Key landings in value, 2012

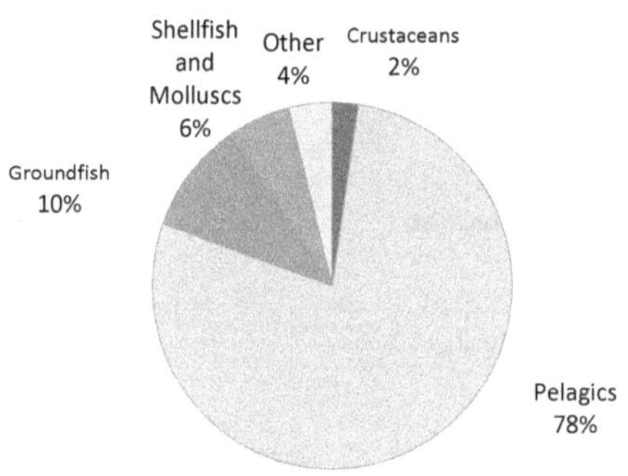

Panel B. Trade evolution

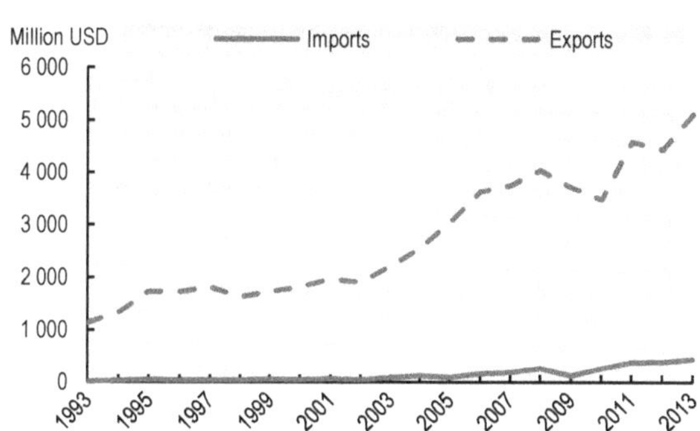

Panel C. Evolution of government financial transfers for marine capture fisheries

USD	Average 2005-10*	2012	% change
Direct payments	7 127 623	14 371 975	102
Cost reducing	..	..	..
General services	54 523 354	75 097 963	38

* Or closest available years.

Panel D. Capacity

	Average 2005-10*	2012	% change
Number of fishers	73 585	93 576	27
Number of fish farmers	17 168	13 854	-19
Total number of vessels	7 724	26 805	247
Total tonnage of the fleet	182 505	160 360	-12

* Or closest available years.

PEOPLE'S REPUBLIC OF CHINA

Summary and recent developments

- Fisheries output showed a slight increase both in capture fisheries and aquaculture in 2012, with aquaculture representing around 73% of the total volume. The growth rate of aquaculture in THE People4s Republic of China (hereafter "China") in 2012 was 6.4%, an impressive amount but still below world average of 7.5%. China remains the largest aquaculture producer in the world. The growth rate of capture fisheries by comparison was 2.5%.
- In 2011 China published the twelfth five-year ocean development plan, covering the years 2011 to 2015. It features plans to optimise the location of ocean industries, restructure traditional ocean industries and other improvements such as rearranging, standardisation and greening of aquaculture farms. China is also working to develop deep sea cage aquaculture and recirculating aquaculture system.
- Official policy is to support and enhance distant water fishing. Distant-fishing enterprises are actively involved in constructing and updating vessels and equipment. Chinese companies have increased opportunities for distance water fishing, including krill fishing in the Antarctic (CCAMLR area) and tuna fishing.

Capture fisheries and aquaculture production

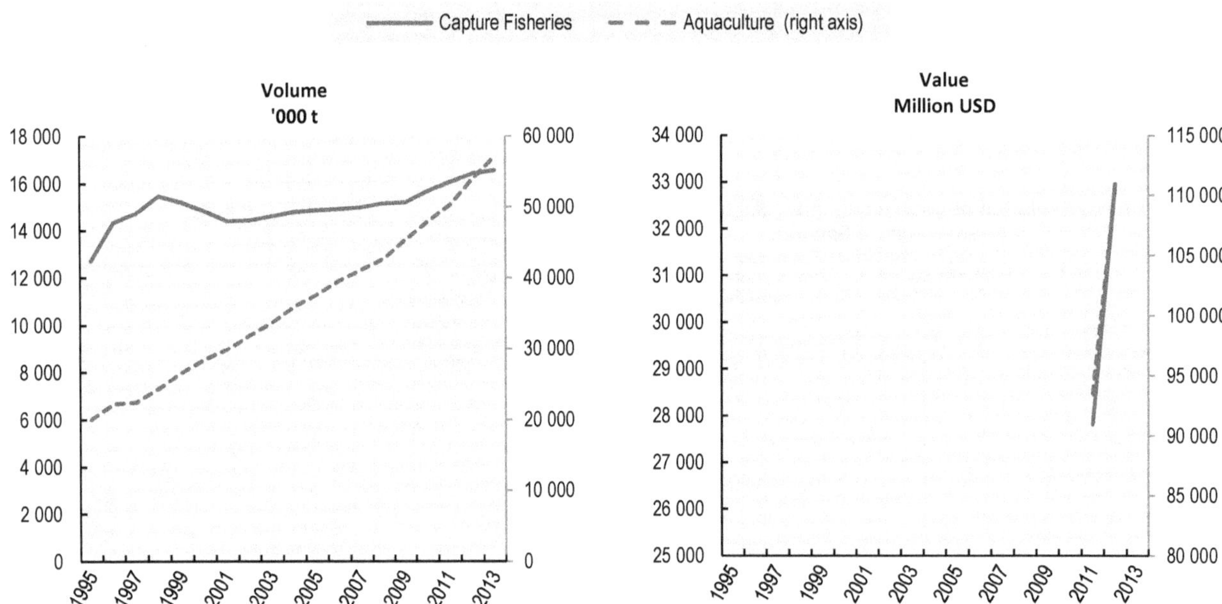

Source: FAO FishStat Database.

Key characteristics of Chinese fisheries

- Total fisheries output in 2012 amounted to around 57 million tonnes, of which about 73% originated from Aquaculture. Inland Aquaculture accounted for 46.1% by volume, followed by Marine Aquaculture (25.6%), Marine Capture (24.2%), and Inland Capture (4%) (Panel A).
- Fish account for 60% of total production by volume. The next most important product categories are shellfish and crustaceans. Carps are by far the most important species for freshwater aquaculture, accounting for 65% of the total volume. Shellfish are the most important marine aquaculture species.
- In 2012, exports of fisheries products totalled USD 18 billion, an increase of 7% over 2011. Major importers of Chinese seafood were Japan, the United States, the European Union, and ASEAN countries. Leading export items were tilapia, shellfish, prawn, and crab. Imports decreased by 2% in value terms. The Russian Federation, the United States, and Peru are the most important exporters to China.
- Data on financial transfers by government are only incompletely available in China. However, a partial picture is available. USD 36 million (RMB 200 million) went to fisheries industry to guarantee supplies of aquatic products in domestic markets. USD 1.8 billion (RMB 8 billion) were invested in modernisation and construction of fishing vessels. Cost reducing transfers included a fuel subsidy amounting to USD 5.7 billion (RMB 35.1 billion) in 2011-12. In the category of general services, USD 65.2 million (RMB 400 million) were provided for restocking exercises, and USD 16.3 million (RMB 100 million) for research in breeding, feed, and fisheries resource sustainability.

Key fisheries indicators

Panel A. Fisheries-products by volume in 2012

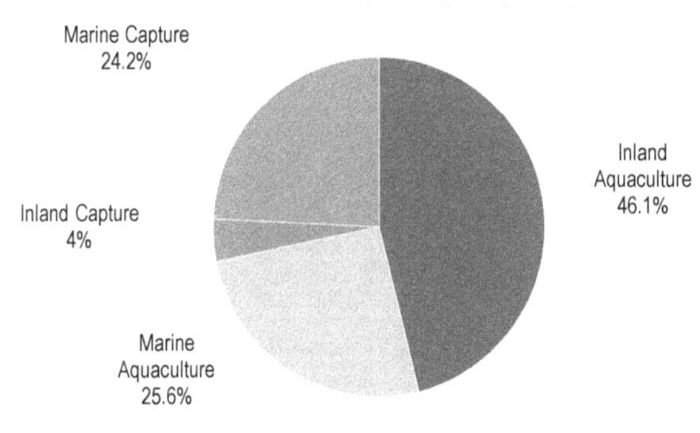

Panel B. Key Species by Volume in 2012

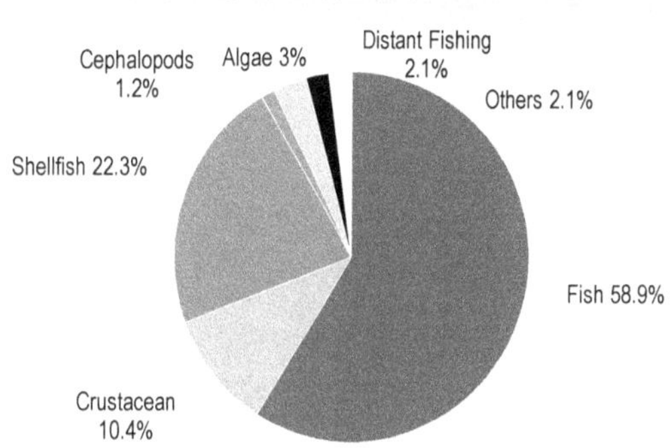

Panel C. Trade evolution (Million USD)

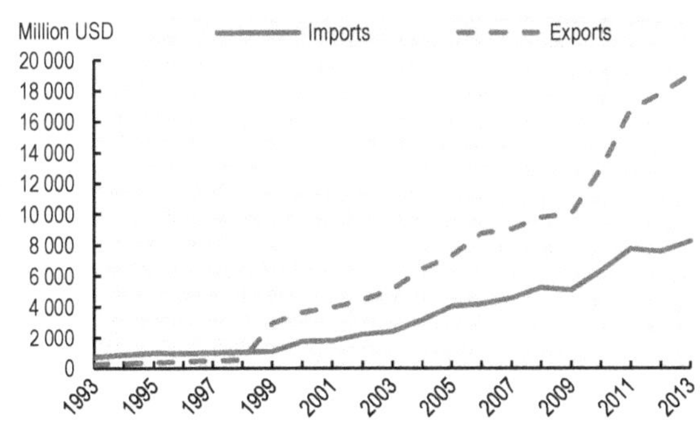

CZECH REPUBLIC

Summary and recent developments

- In the Czech Republic, financial support provided by the European Maritime and Fisheries Fund (EMFF) will focus on the further development of a competitive and environmentally friendly fish farming sector.
- Fish consumption per capita was estimated at 6.5 kg in 2012, which is lower than the average of around 11 kg per year in Europe. Marine fish represent the majority of consumption while domestic production is from freshwater capture or aquaculture.
- Domestic and export markets tend to be very traditional and stable, mainly composed of live carp. Production is matched to demand to maintain stability in the market.

Aquaculture production

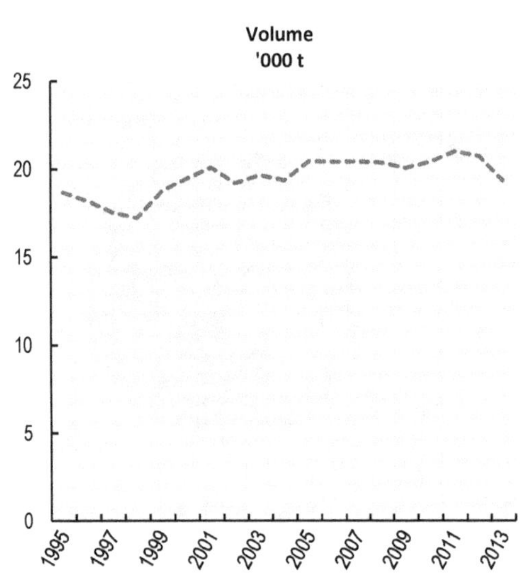

Source: FAO FishStat Database.

Key characteristics of Czech Republic fisheries

- Production of fish in the Czech Republic is almost exclusively from aquaculture, with carp (84%) being by far the most important species, followed by rainbow trout (4%). Carp production in 2013 was 16 809 tonnes out of a total production of 20 763 tonnes. (Panel A)
- The Czech Republic, being a landlocked country, is a net importer. Imports were more than double exports in 2012, totalling 44 311 tonnes while exports were 19 363 tonnes. Imports consist mostly of marine species in fresh, frozen and processed form. In line with the national production patterns, carp is the most exported species. (Panel B)
- Government Financial Transfers are provided only to the aquaculture sector, as there is no marine fishery. Total Government Financial Transfers were in 2013 were CZK 392 million, more than 90% of which was for the removal of pond silt. (Panel C)
- The number of aquaculture operations and their employed staff is very stable. In 2012 the Czech Fish Farmers Association had 71 members. (Panel D)

Key fisheries indicators

Panel A. Key aquaculture species in value, 2013

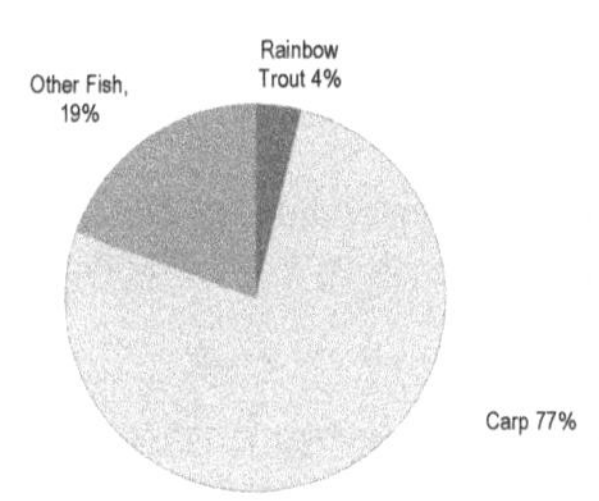

Panel B. Trade evolution

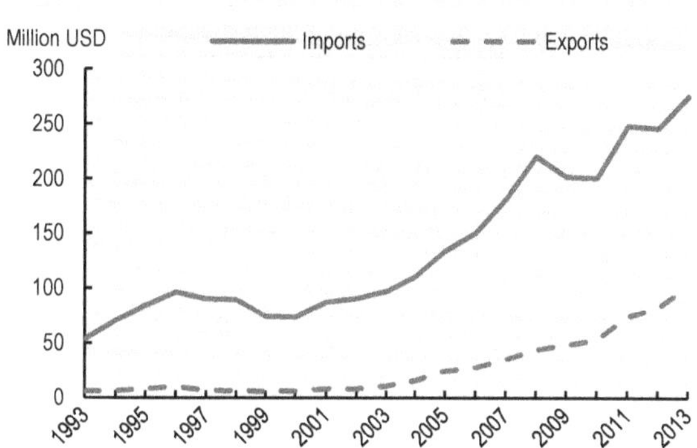

Panel C. Evolution of government financial transfers for aquaculture

USD	Average 2005-10*	2013	% change
Direct payments	20 432 621	20 024 872	-2
Cost reducing	..	..	..
General services	1 852 338	..	..

* Or closest available years.

Panel D. Capacity

	Average 2005-10*	2013	% change
Number of fishers	..	..	..
Number of fish farmers	1 463	1 428	-2
Total number of vessels	..	..	..
Total tonnage of the fleet	..	..	..

* Or closest available years.

DENMARK

Summary and recent developments

- During 2014 Denmark expects to submit fisheries management measures for the protection of reef structures in 10 Natura 2000 sites in Kattegat and the Western Baltic Sea to the European Commission. These measures are to be discussed with relevant member states having a direct management interest in the areas before being submitted to the EU Commission as Joint Recommendations in accordance with the new basic regulation under the Common Fisheries Policy.

- During 2008-2014 Denmark conducted trials with catch quota management with full documentation (CQM). In the trials all catches of cod were counted against the quota and monitored by CCTV. The experience from the trials will assist Denmark in implementing the landing obligation under the reformed Common Fisheries Policy and should result in a more sustainable fishery while simplify existing regulations.

- The Danish Government will in 2014 adopt a National Strategic Plan for Development of Sustainable Aquaculture 2014-20. The strategic plan will contain a target for the increase in the production of fish and shellfish and for the increase in the export of feed, feed ingredients and technology for the aquaculture sector.

- A programme for implementing the European Marine and Fisheries Fund 2014-2020 in Denmark will be submitted to the European Commission in 2014. The new programme will focus on sustainable fisheries and aquaculture.

Capture fisheries and aquaculture production

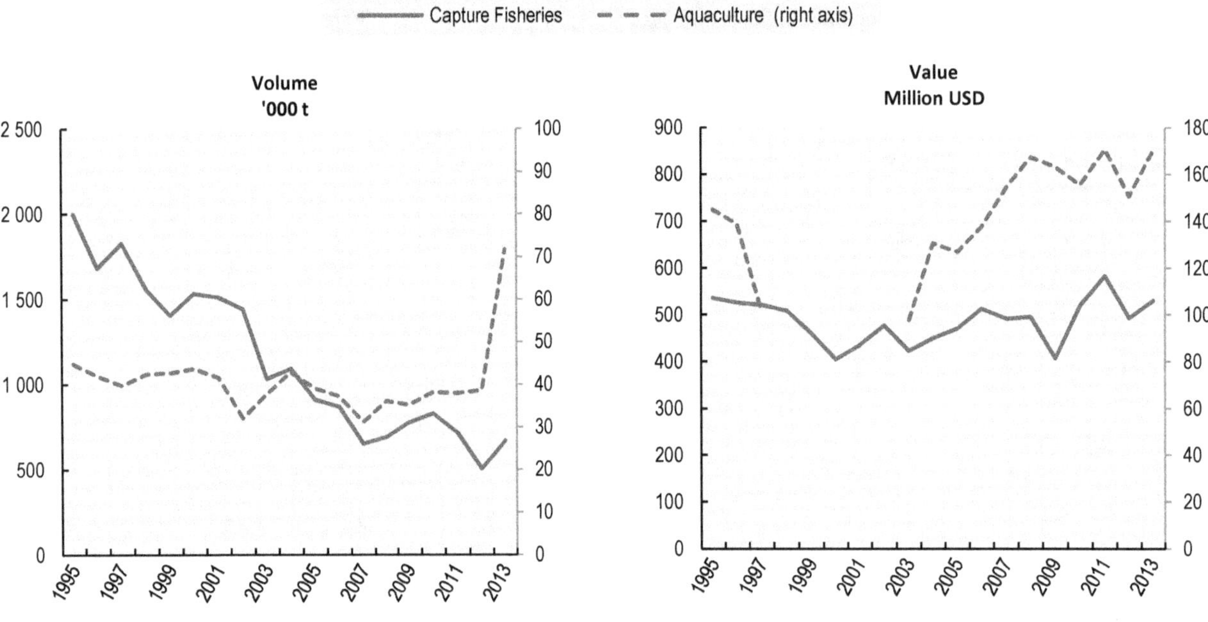

Source: FAO FishStat Database.

Key characteristics of Danish fisheries

- The mainstay of Danish fisheries consists of pelagics (herring and mackerel and species for reduction to fish meal and oil), groundfish (e.g. cod) and shrimps. (Panel A)
- Exports consist of several species, including salmon, groundfish, shrimps, herring, flatfish, fishmeal and oil. Trade in salmon has been increasing, whereas trade in groundfish has decreased. Other EU countries purchase 72% of Danish exports value (2010), while exports to other parts of the world, including central and Eastern Europe and China, are increasing. (Panel B)
- In 2013, DKK 188.2 million, financed evenly by the EU and Danish public funds, were provided in financial assistance to the sector. National support schemes include a general measure to encourage development, sustainability and innovation in the food industry sector. In addition, the government pays for management, control and research into capture fisheries and aquaculture amounting to approximately DKK 399 million in 2012. (Panel C).
- The number of commercially active vessels in the Danish fleet fell 21% over the period 2007-2012. Employment also fell substantially. The economic performance for active vessels peaked in 2011. A new model of regulation with transferable fishing concessions (TFC) has had important positive effects on the economic performance of the fleet. (Panel D) Four hundred and twenty-five people are directly employed in Danish aquaculture, mainly in traditional fish farming. Aquaculture production in Denmark is mainly focused on rainbow trout.

Key fisheries indicators

Panel A. Key landings in value, 2013

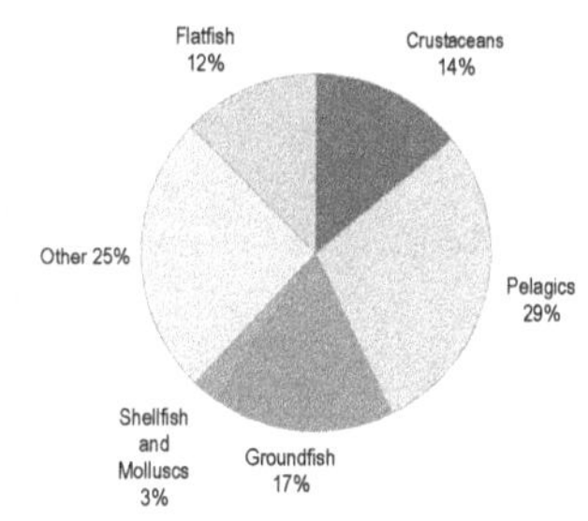

Panel B. Trade evolution

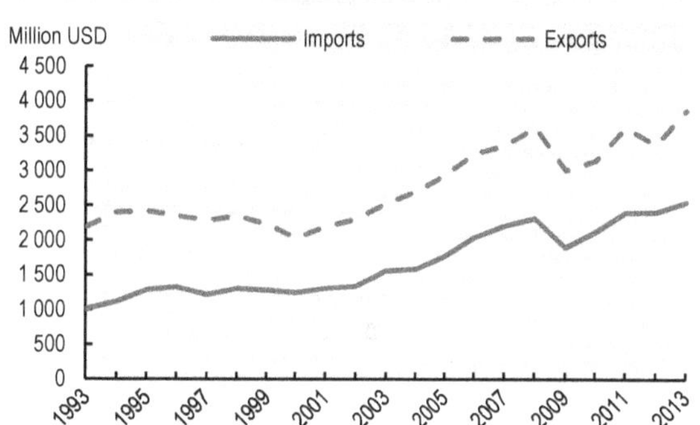

Panel C. Evolution of government financial transfers for marine capture fisheries

USD	Average 2005-10*	2013	% change
Direct payments	9 993 185	2 629 321	-74
Cost reducing	847 613	45 523 353	5271
General services	68 035 348	72 079 818	6

* Or closest available years.

Panel D. Capacity

	Average 2005-10*	2013	% change
Number of fishers	2 511	1 891	-25
Number of fish farmers	509	421	-17
Total number of vessels	2 996	2 663	-11
Total tonnage of the fleet	76 823	65 386	-15

* Or closest available years.

ESTONIA

Summary and recent developments

- The main goal as stated in the Estonian Fisheries Strategy 2014-20 is to develop Estonian fisheries as sustainable branch of economy by raising the competitiveness of fisheries products in internal and external markets.
- Policies will target adding value to the pelagic fish stocks caught in Baltic Sea in order to increase the use of Baltic herring and sprat for human consumption.
- Successful control of effort will allow fishing seasons to be lengthened where markets demand a more stable flow of fresh fish. Maintaining a balance between capacity and fishing opportunities is key to achieving this.
- A new processing plant has been recently built for processing aquaculture products to improve their market position in key markets in Ukraine and Russia which demand large volumes. However, recent developments in these countries could impact export demand in the near term.

Capture fisheries and aquaculture production

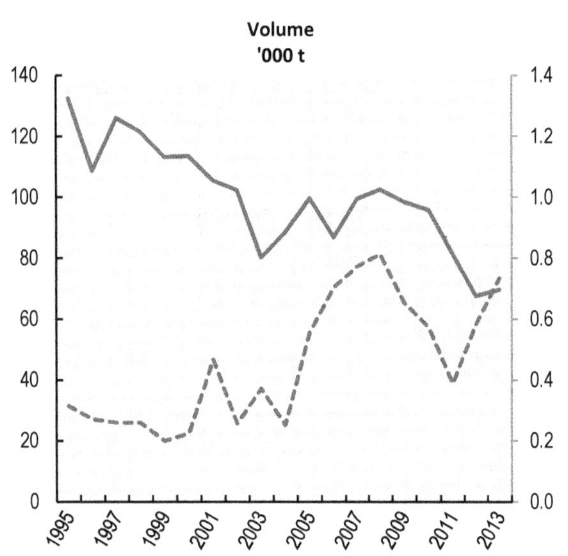

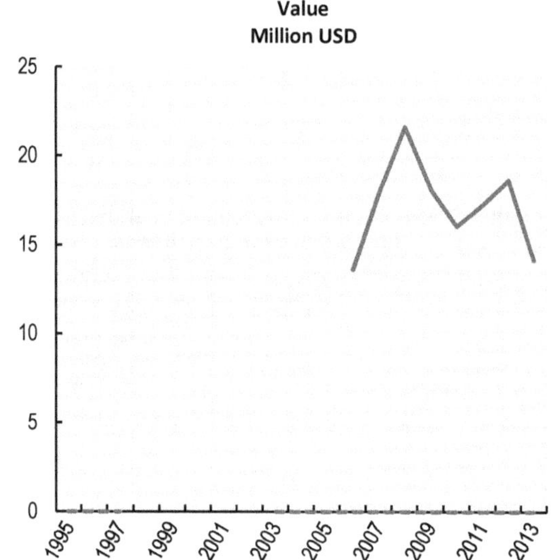

Source: FAO FishStat Database.

Key characteristics of Estonian fisheries

- In 2012 total landings of the Baltic Sea coastal and open sea, inland water bodies and Atlantic Ocean were 67 171 tonnes and EUR 52.27 million (USD 67.12 million) by value. Pelagic fisheries predominate, with the main species being Baltic herring and sprat, followed by cod and perch in the Baltic and Northern prawn and redfish in the high seas. (Panel A)
- The fishing sector is a net exporter, and the importance of trade in total production has been growing in recent years. Export value in 2012 was EUR 176.3 million (USD 226.39 million), while imports were valued at EUR 111.9 million (USD 143.69 million). (Panel B)
- Total value of Government Financial Transfers in 2012 was EUR 15.1 million (USD 19.39 million), the majority of which coming from the European Fisheries Fund (EFF). Of the total, the majority was provided for Axis 4 projects—"Sustainable development of fisheries areas" (Panel C).
- Consolidation of the fisheries sector continues, with a 22% reduction in fleet capacity between 2005-10 and 2012, though some of this reduction is likely the result of demand reduction following the financial crisis. This demand decline is also seen in the reduction of the number of employed in aquaculture, which is contrary to recent trends. In 2012, 1 569 people were employed in marine coastal fishing, 474 in marine deep-sea fishing and 551 in the inland waterbodies harvesting sector. (Panel D)

Key fisheries indicators

Panel A. Key landings in value, 2012

Panel B. Trade evolution

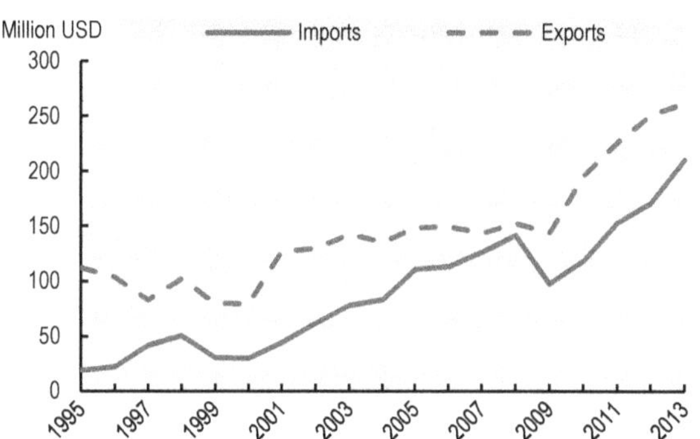

Panel C. Evolution of government financial transfers for marine capture fisheries

USD	Average 2005-10*	2013	% change
Direct payments	6 190 096	14 932 240	141
Cost reducing	217 907	382 329	75
General services	7 626 078	12 665 164	66

* Or closest available years.

Panel D. Capacity

	Average 2005-10*	2013	% change
Number of fishers	3 554	2 594	-27
Number of fish farmers	98	84	-14
Total number of vessels	1 014	1 460	44
Total tonnage of the fleet	19 018	15 217	-20

* Or closest available years.

FRANCE

Summary and recent developments

- Responsibility for management of fisheries and aquaculture has been transferred from the Ministry for Agriculture and Fisheries to that for Ecology, sustainable development and energy since 2012.
- From 2007 to 2013, EUR 216 million were allocated to France through the European Fisheries Fund (EFF). With additional support by the French government, the sector benefited from over EUR 450 million for modernising and adapting the sector. EUR 558 million were allocated to France in the new European maritime and fisheries fund (EMFF) for the 2014-20 period.
- Regional spatial planning schemes for the development of marine aquaculture have been elaborated. They locate existing production sites and identify suitable potential new sites.
- Since 2012, three marine protected areas ("parcs naturels marins", PNM) have been created (PNM des Glorieuses, 43 000 km², Indian Ocean ; PNM des estuaires picards et de la mer d'Opale; PNM du Bassin d'Arcachon).
- A public eco-label is in the process of being created, following a 2012 law decree. A commission bringing together sector representatives as well as NGOs, consumers and independent experts has validated a series of tractability, social and quality criteria. They will be publicly discussed in the course of 2014.
- The Patagonian toothfish fishery of Kerguelen EEZ has been certified by "MSC".

Capture fisheries and aquaculture production

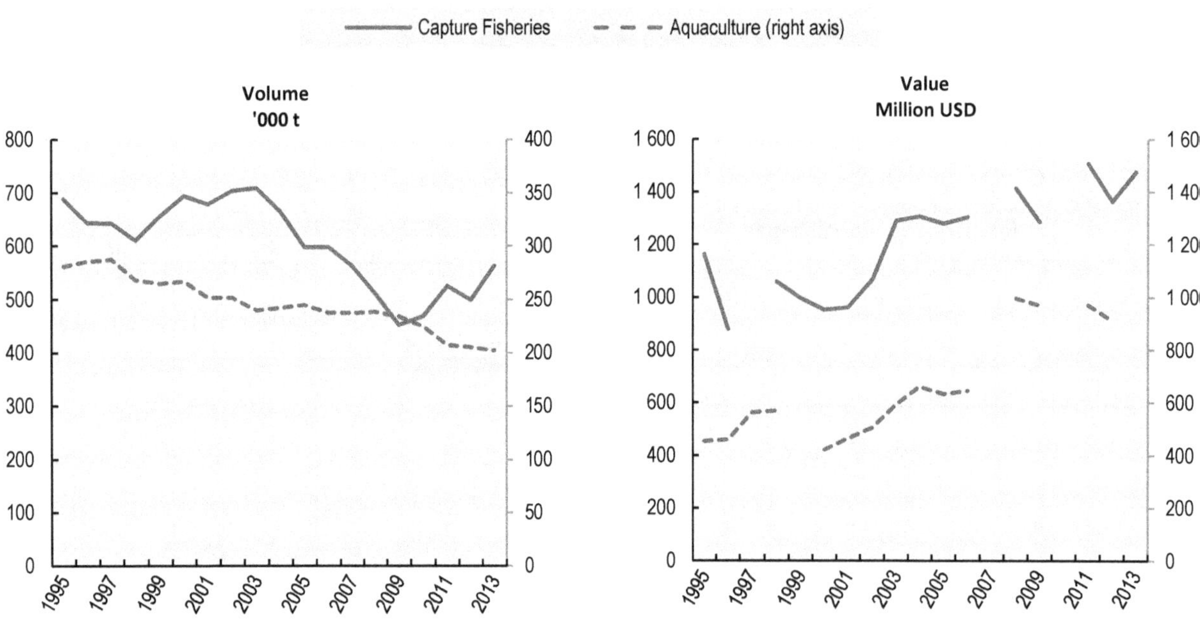

Source: FAO FishStat Database.

Key characteristics of French fisheries

- In 2013, 533 thousand tonnes of fishery products were harvested by the French fleet, for a value of EUR 1.1 billion (USD 1.5 billion). This represents about 10% of EU captures in volume, of which 80% originate in the NE Atlantic. Captures increased by about 3% in volume and 4% in value since 2012.
- In 2013, imports of fishery products reached 1.1 million tonnes and almost EUR 5 billion (USD 6.6 billion), while exports reached about 325.000 tonnes and EUR 1.4 billion (USD 1.9 billion).
- The French fleet employed about 22 000 fishers 2012, about 3% down from 2010, while first-hand sale and transformation sector accounted for another 21 000 jobs.
- In 2013, the French fleet counted more than 7 000 vessels. The fleet has decreased by a bit more than 10% since 2011.

Key fisheries indicators

Panel A. Key landings in value, 2013

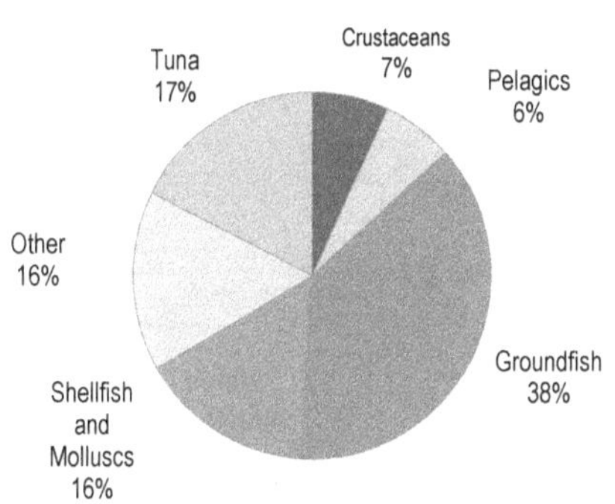

Panel B. Trade evolution

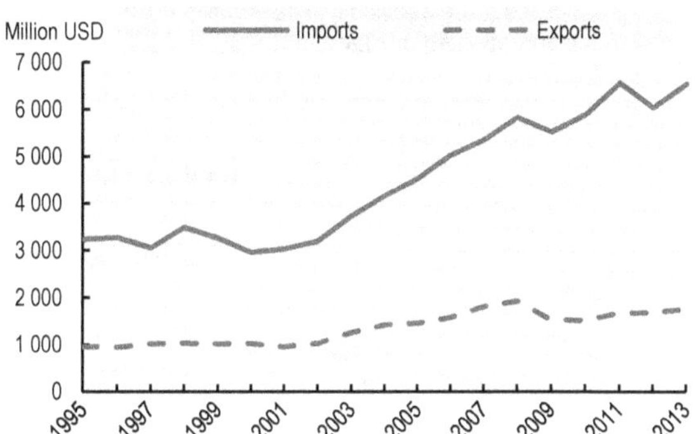

Panel C. Evolution of government financial transfers for marine capture fisheries

USD	Average 2005-10*	2013	% change
Direct payments	54 956 458	..	..
Cost reducing	163 994 969	..	..
General services	36 599 942	..	..

* Or closest available years.

Panel D. Capacity

	Average 2005-10*	2013	% change
Number of fishers	19 554	16 887	-14
Number of fish farmers	20 349	20 125	-1
Total number of vessels	7 605	7 158	-6
Total tonnage of the fleet	198 814	166 561	-16

* Or closest available years.

GERMANY

Summary and recent developments

- Since 2012, Germany's fishing fleet has decreased by approximately 50 fishing vessels most of which were from small-scale coastal fisheries. The capacity and the number of vessels have by and large remained constant in the high sea fisheries sector and in respect of large cutters with a length of 18+ metres.
- Landings by German fishing vessels in German and foreign ports totalled 209 600 t in 2013 with a revenue of EUR 190.8 million, 10% above the previous year.
- Approximately 25 000 tonnes of fish and shellfish with a value of at least EUR 80 million were produced in German aquaculture facilities. The share of aquaculture fish products in the German market grew steadily to 27 % of total domestic fish demand in 2013.

Capture fisheries and aquaculture production

Capture Fisheries — Aquaculture (right axis)

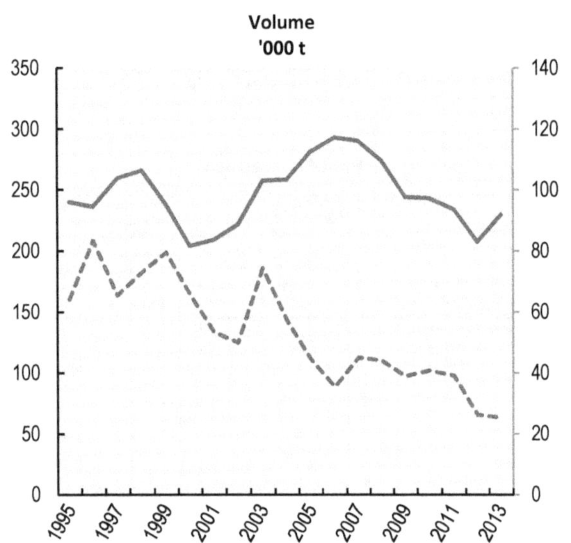

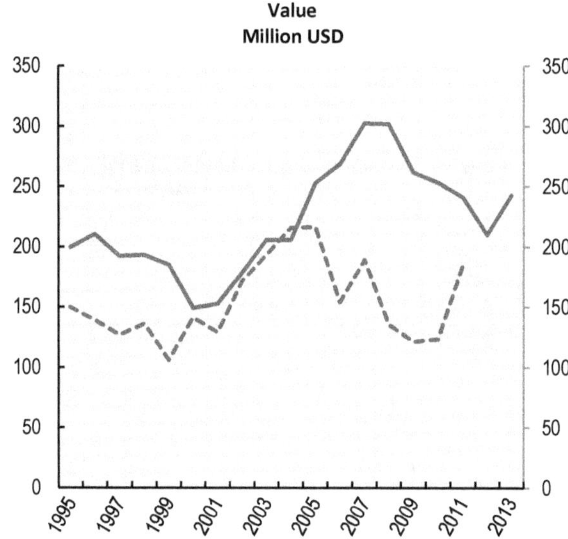

Source: FAO FishStat Database.

Key characteristics of German fisheries

- The prices of the economically important fish i.e. herring, saithe, cod and plaice, were somewhat lower than in 2012. This was particularly true of plaice fishing, where the low prices kept revenue low despite good stock conditions and good catches. Shrimp fishers were able to achieve stable prices for the second year in a row (Panel A).
- In 2013, 870 000 tonnes of fish were imported into Germany and 515 000 tonnes were exported. Poland and Denmark were the most important EU suppliers of fish to Germany. However, the People's Republic of China was the single biggest supplier of fish. Frozen fish accounted for around 35% of total consumption while canned fish and marinades make up just under one-third of all fishery products. Fresh fish (trending downward), smoked fish, fish salads and other fishery products making up the rest (Panel B).
- Expanded consumer information and traceability of fishery products required investments to be made by the Germany fisheries sector in scanners, coding devices and the related software. By the end of 2013, five German enterprises had invested around EUR 400 000 in this technology. They were reimbursed 90% of the costs from EU funds. Three other projects, with a support volume of around EUR 500 000 have already been promised by the EU.
- The German fishing fleet currently consists of 1 530 units with a total tonnage capacity of 61 061 GRT (gross register tonnes) and an engine power of 142 751 kilowatts (Panel D).

Key fisheries indicators

Panel A. Key landings in value, 2013

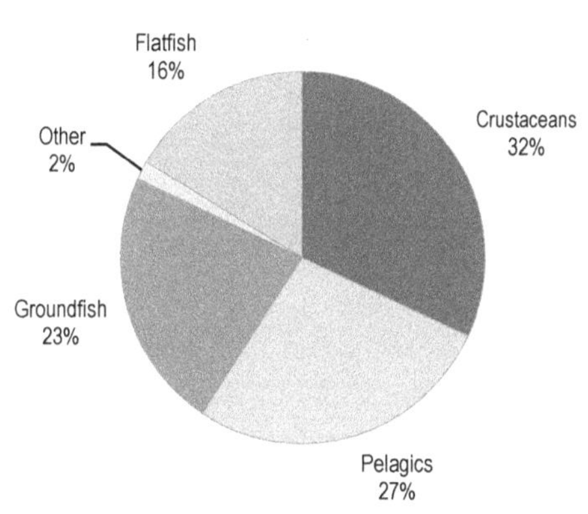

Panel B. Trade evolution

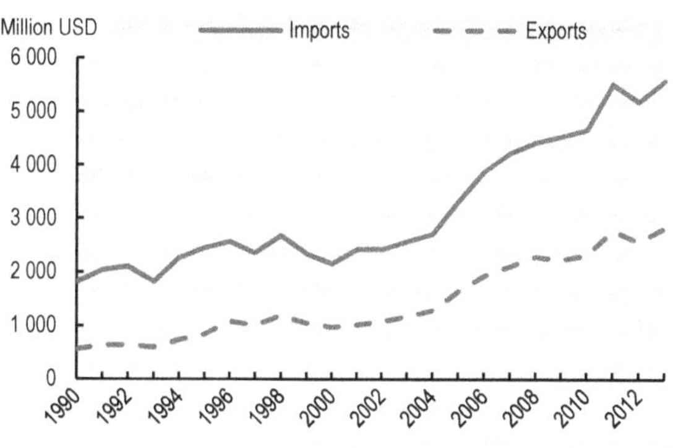

Panel C. Evolution of government financial transfers for marine capture fisheries

USD	Average 2005-10*	2013	% change
Direct payments	2 003 541	2 837 254	42
Cost reducing	695 393	..	..
General services	5 198 788	814 717	-84

* Or closest available years.

Panel D. Capacity

	Average 2005-10*	2012*	% change
Number of fishers	2 005	4 917	145
Number of fish farmers	..	0	..
Total number of vessels	1 961	1 582	-19
Total tonnage of the fleet	66 434	64 835	-2

* Or closest available years. 2013 for fishers, 2011 for vessels.

GREECE

Summary and recent developments

- In August 2012, administration of the fishery sector was transferred back to the Ministry of Rural Development and Food.
- In 2013, the OECD was asked to estimate and suggest ways to reduce administrative burdens in 13 sectors of the Greek economy among which fisheries. The recommendations, including setting licensing at one-stop-shop and designing producer-seller licenses, are in the process of being implemented.
- The obligation for trawlers and purse seines to deliver the catch in the nearest public auction hall possible was modified. Trawlers and purse seines may now deliver their catch to the biggest public auction halls such as Athens, Thessaloniki, Kavala or Patra.
- The setting up of an Electronic reporting System for fishing activities is being implemented. The system should be fully operating next year.

Capture fisheries and aquaculture production

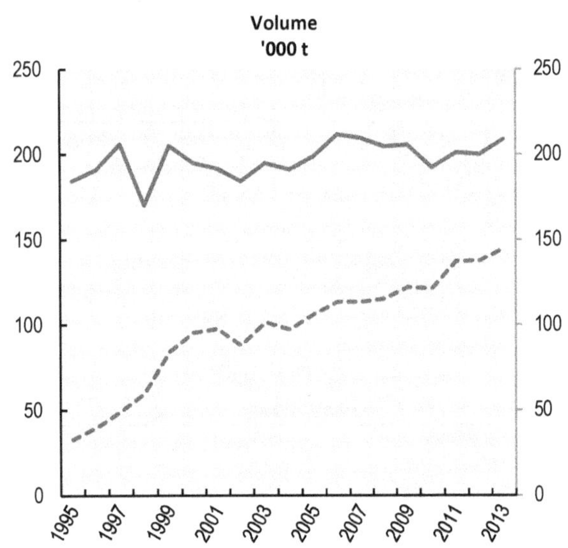

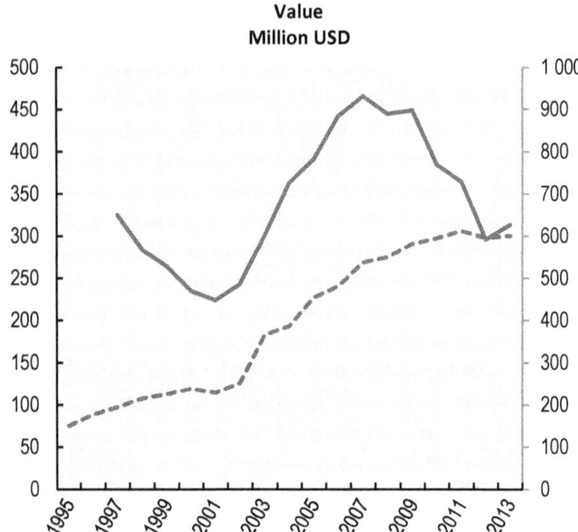

Source: FAO FishStat Database.

Key characteristics of Greek fisheries

- The Greek fishing fleet is the largest in vessel number in the European Union due to a high number of small vessels but it represents only 4.5% of the Community fleet in terms of capacity. At the end of 2013, the fleet counted over 15 000 vessels, of which more than 95% are coastal vessels, with a total capacity of over 78 000 GT.
- Total employment in the marine capture fisheries sector amounted to about 36 000 persons in 2012 and 2013 (with about 6 000 operating on deep sea vessels, and 30 000 on coastal vessels)
- In 2012, Greek aquaculture production reached about 114 000 tonnes for a value of EUR 464 million (USD 616 million). About 82% of this production volume and about 96% of the value derives from marine finfish aquaculture (mostly seabream and seabass), with the remaining mainly coming from shellfish products. In 2012, the total number of aquaculture farms both in marine and inland waters in Greece reached 1 052 units.

Key fisheries indicators

Panel A. Key landings in value, 2013

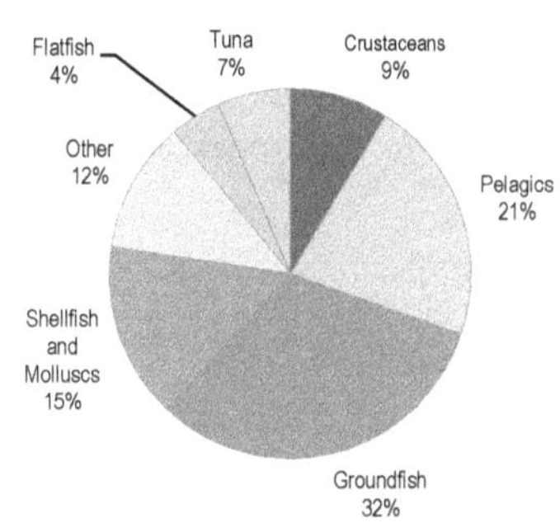

Panel B. Trade evolution

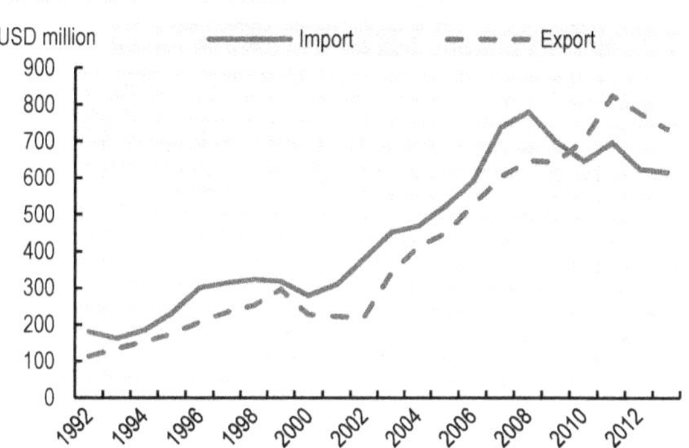

Panel C. Evolution of government financial transfers for marine capture fisheries

USD	Average 2005-10*	2013	% change
Direct payments	19 886 158	15 514 031	-22
Cost reducing	34 087 459	..	..
General services	9 298 870	10 026 533	8

* Or closest available years.

Panel D. Capacity

	Average 2005-10*	2013	% change
Number of fishers	33 459	35 976	8
Number of fish farmers	6 116	4 796	-22
Total number of vessels	17 736	16 249	-8
Total tonnage of the fleet	90 330	80 783	-11

* Or closest available years.

ICELAND

Summary and recent developments

- The present Icelandic management system is built on individual transferrable quotas (ITQ) issued to vessels. The Minister for Fisheries receives advice from the Marine Research Institute (MRI) and consequently issues total allowable catch (TAC) for individual stocks for the fishing year, which runs from 1 September to 31 August the following year. The size of each vessels annual catch quota for each stock is its share in the stock multiplied by the TAC of that stock. Both the permanent quota-shares and the annual catch quotas are transferable in part or total, subject to certain restrictions.
- The share of marine fisheries products in Icelandic GDP and export earnings are significantly higher than in most other OECD countries, making the Icelandic economy one of the most fisheries-dependent in the world. The share of fisheries, including fishing and fish processing has historically been over 10%, with a "dip" during the years just before the financial crisis, when the Icelandic krona appreciated sharply.
- In 2012-13, there was a noticeable transfer of workforce from fishing sector to fish processing. More fish is landed unprocessed and is now processed in Iceland.

Capture fisheries and aquaculture production

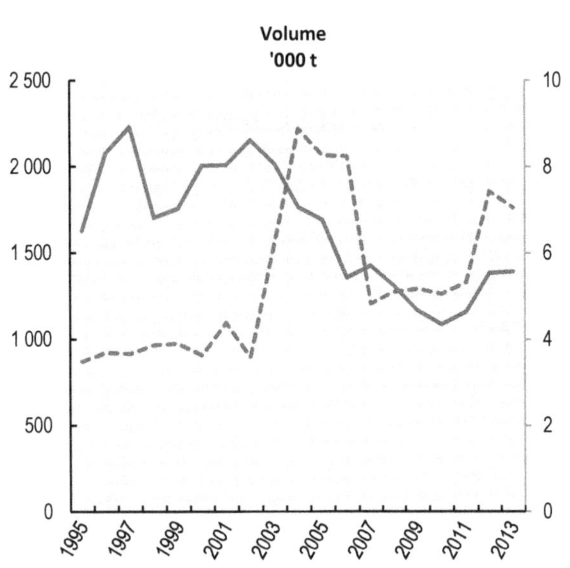

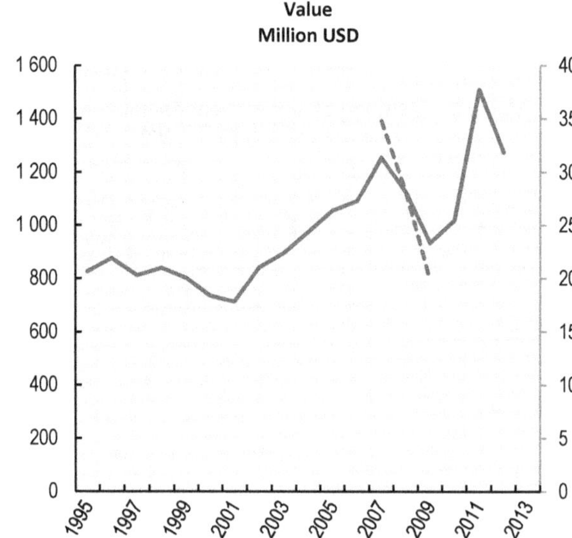

Source: FAO FishStat Database.

Key characteristics of Icelandic fisheries

- Icelandic marine landings in 2013 amounted to 1 362 000 tonnes with the bulk of landings from small pelagic species such as capelin, herring and mackerel. The years 2012 and 2013 are among the most profitable in the fishing industry, with the profits largest in the pelagic sector reflecting a high world market price of fish meal and oil, but the demersal sector is also enjoying high profit figures (Panel A).
- In 2013 the export value of marine fisheries products amounted to ISK 272 billion. The largest share of marine fisheries products are exported to the European Union, with the United Kingdom being the single largest market. Cod remains the most valuable export species around 30% of the total fisheries export value, followed by capelin with around 12% share (Panel B).
- There are no government sponsored decommissioning schemes or direct payments to the fishing or fish processing sectors in Iceland. Fuel for the fishing fleet, as well as other vehicles that do not use the road system is exempt from a "road fee". In 2013 a tax concession for crew members of fishing vessels dependent on number of days at sea for each individual was discontinued after several years of phasing out (Panel C).
- The Icelandic fishing fleet consisted of 1 696 vessels at the end of year 2013. There are three main categories of vessels, undecked vessels, decked vessels of various sizes and trawlers. The small vessels or undecked vessels as they are officially classified are on average 5 GT in size and target demersal species, mainly cod (Panel D). There was an increase in number of small vessels but a steady decrease in larger vessels. This might be connected to the coastal fisheries common pool quota for demersal species adopted in 2009.

Key fisheries indicators

Panel A. Key species by value in 2012

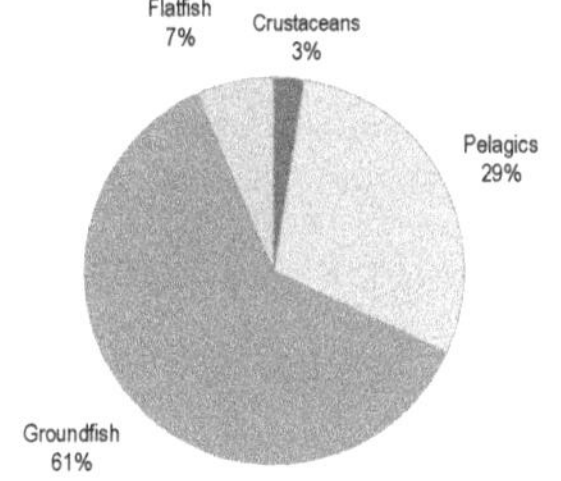

Panel B. Trade evolution

Panel C. Evolution of government financial transfers

USD	Average 2005-10*	2012	% change
Direct payments	..	..	..
Cost reducing	12 066 657	7 872 568	-35
General services	34 634 355	25 647 788	-26

* Or closest available years.

Panel D. Capacity

	Average 2005-10*	2013	% change
Number of fishers	4 558	3 600	-21
Number of fish farmers	200	..	..
Total number of vessels	1 314	1 394	6
Total tonnage of the fleet	155 545	213 561	37

* Or closest available years.

INDONESIA

Summary and recent developments

- According to The State of World Fisheries and Aquaculture - FAO 2014, Indonesia is the second largest producer of marine capture fisheries and the second largest producer of aquaculture, and has the second highest number of fishers and aquaculture farmers.
- The new Law No. 32 of 2014 concerning the Ocean stipulated that the central and local governments should work together using the principles of the blue economy to maximise public benefits of marine resources, including fisheries, energy and mineral resources, coastal resources and small islands as well as unconventional resources.
- In cooperation with FAO, Indonesia has been developing a pilot project of an aquaculture-based blue economy approach in Nusa Tenggara Barat province, focusing in Central Lombok and East Lombok Districts and bases on the principles of sustainability, nature's efficiency, zero waste, and social inclusiveness.
- Indonesia has actively participated in the Regional Fisheries Management Organizations (RFMOs) where has been a member of the Indian Ocean Tuna Commission (IOTC) since 2007, a member of the Commission for the Conservation of Southern Bluefin Tuna (CCSBT) since 2008, and a member of the Western and Central Pacific Fisheries Commission (WCPFC) since 2013.
- The National Medium Term Development Plan (RPJMN) 2015 – 2019 calls for development of the marine and fisheries sectors within five years focussing on three pillars: sovereignty, sustainability, and prosperity. These pillars form the basis of the vision of the "realization of management of marine resources and fisheries by sovereign, independent and sustainable prosperity for the people."

Capture fisheries and aquaculture production **Trade evolution**

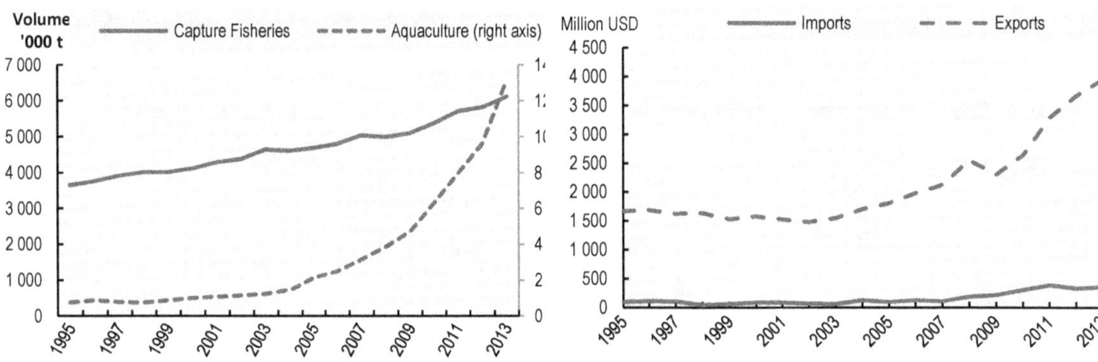

Source: FAO statistics.

Source: UN Comtrade database.

IRELAND

Summary and recent developments

- In 2011, landings of fish by Irish registered vessels totalled 202 819 tonnes with a total value of EUR 265.7 million. In 2012 the total volume was 273 861 tonnes with a corresponding value of EUR 262.3 million.
- Aquaculture production volume decreased from 47 407 tonnes in 2009 to 36 384 tonnes in 2012 and 34 665.7 tonnes in 2013. Unit value in the oyster and salmon sectors enabled overall production value to increase despite the volume decrease, to a high of EUR 130.8 million in 2012. Thereafter overall production value was negatively affected by the continuing drop in production volume, worth EUR 117.9 million, in 2013.
- Over the period 2010-13, there has been further consolidation in the processing sector, with an increasing focus on value-adding and maximising returns on limited fish raw material. The catching sector has focussed on increasing quality, on board processing in tandem with delivering certified responsible fishing practices.
- In 2013 Irish seafood exports were valued at almost EUR 489 million and amounted to 254 891 tonnes. This represents a 7% decrease in value from EUR 525 million in 2012 with a corresponding 12% decrease in volume from 289 860 tonnes in the same year.

Capture fisheries and aquaculture production

Capture Fisheries — Aquaculture (right axis)

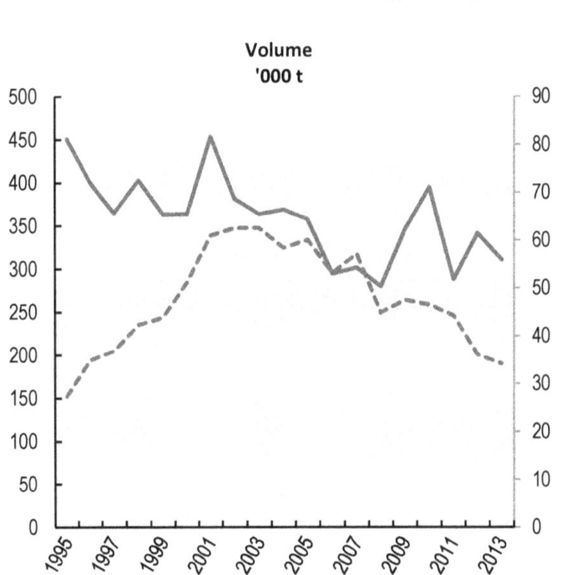

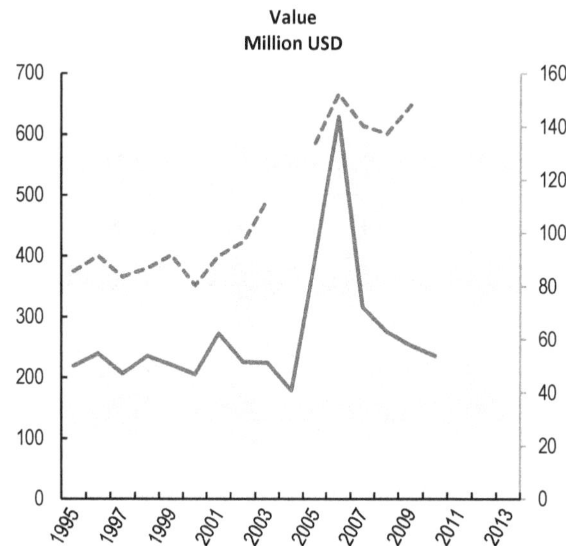

Source: FAO FishStat Database.

Key characteristics of Irish fisheries

- The most important species landed in Ireland in 2012 in terms of value were pelagic fish (38%), groundfish (26%), crustaceans (19%), shellfish and molluscs (8%), flatfish (7%) and tuna (2%) (Panel A).
- Ireland has been making trade surplus in fisheries products over decades. The total export value of fisheries products was EUR 489 million in 2013, being exported mainly to France, Spain, the United Kingdom and Italy. The total import value of fisheries products amounted to EUR 236 million in 2013 resulting in trade surplus of EUR 253 million in this year (Panel B).
- The amount of GFTs paid in 2013 amounted to EUR 4.249 million, down 15% on the previous year. While grants to fleets and aquaculture increased those to the processing sector decreased (Panel C).
- The total numbers of fishers and fish farmers have decreased compared to the year 2008. In total, approximately 11 000 people are employed directly in the sea fishing, aquaculture and support industries of which the employment in aquaculture accounts for 1 822 persons in 2013. The number of vessels increased (+6.89%) and the total tonnage of the fleet decreased (-8.51%) in the same period (Panel D).

Key fisheries indicators

Panel A. Key species by value in 2010

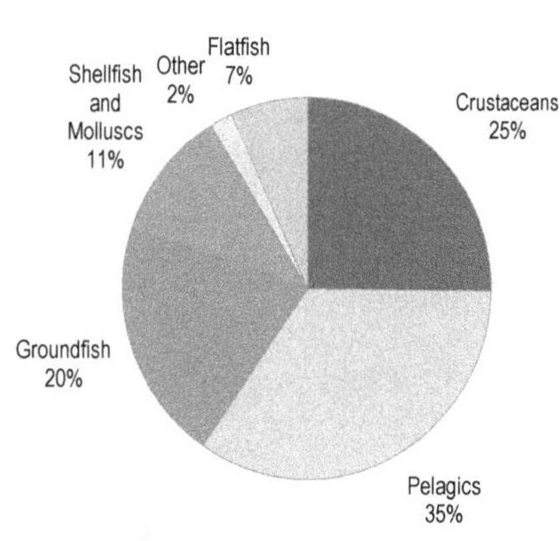

Panel B. Trade evolution

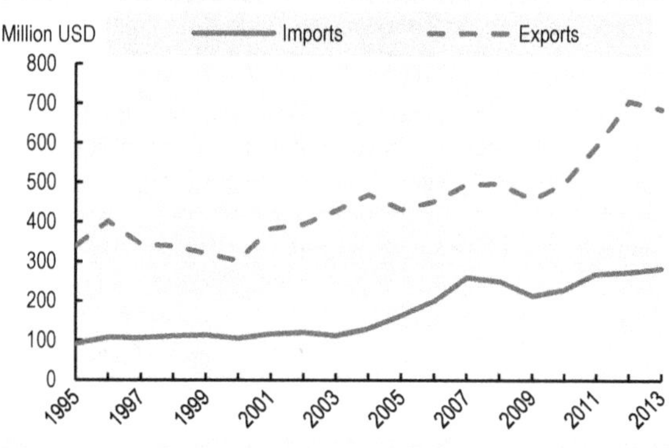

Panel C. Evolution of government financial transfers

USD	Average 2005-10*	2012	% change
Direct payments	19 022 257	..	..
Cost reducing	5 103 089	..	..
General services	194 561 311	..	..

* Or closest available years.

Panel D. Capacity

	Average 2005-10*	2012	% change
Number of fishers	4 448	..	..
Number of fish farmers	1 895	..	..
Total number of vessels	1 934	2 216	15
Total tonnage of the fleet	73 748	64 796	-12

* Or closest available years.

JAPAN

Summary and recent developments

- Fisheries management in Japan has been stable in recent years. The TAC system established in 1996 and the TAE (Total Allowable Effort) system in 2003 are the main tools for Japanese output controls.
- During the last decade, there has been a decreasing trend in fisheries overall. In 2012, the marine aquaculture production shows a rebound after a sharp drop due to the Great Earthquake in 2011. The marine capture fisheries sector was still experiencing reduction in production in 2012, down from 2011 level.
- The basic plan of Japan's fishery policy was renewed in March 2012. In accordance with this plan, "Resource management Plan" was launched. Under this system, the national and prefectural governments formulate resource-management policies, and based on policies, fisher's organisations create resource-management plans and implement them.
- To address decreasing fish consumption, the "Delight of a Fish-Rich Country" project commenced in 2012. It is a campaign as public-private collaboration with overall fishing industry such as fishing operators, processors, distributers, administrations, etc., for promotion to expand consumption of fish and fishery products.

Capture fisheries and aquaculture production

Capture Fisheries — Aquaculture (right axis)

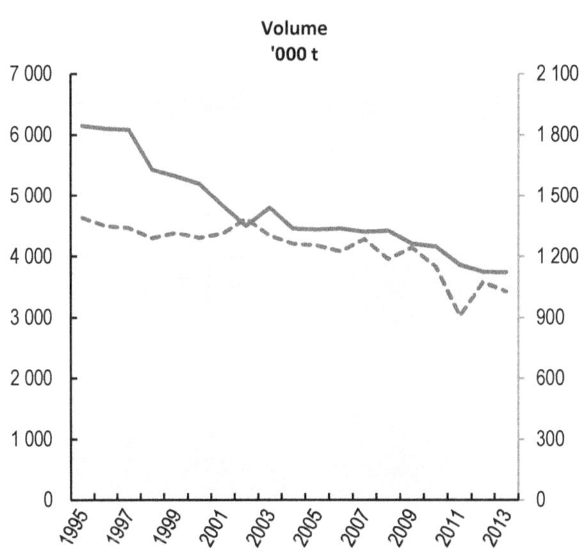

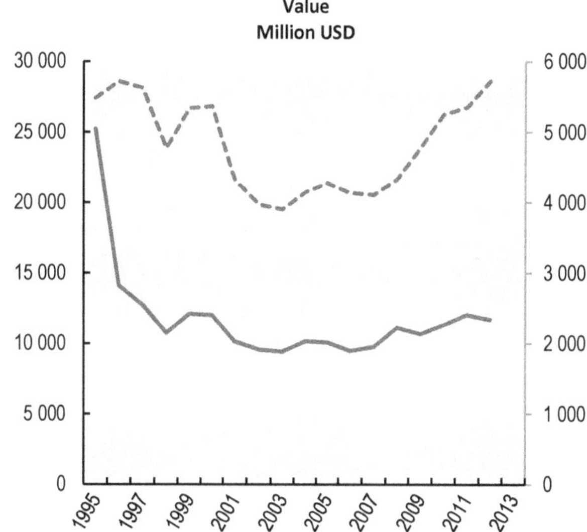

Source: FAO FishStat Database.

Key characteristics of Japanese fisheries

- The main species harvested in Japan are tuna (22%), followed by pelagic species (21%) such as salmon, mackerel, and sardine, shellfish and molluscs (19%), groundfish (8%) and crustaceans (5%) (Panel A).
- Japan is the one of the largest importer of fisheries products. Import value of the fish and fishery products in 2013 decreased from previous year. After the Great Earthquake, the value of imports slightly rose, then, has been falling again due to decreased fish consumption. People's Republic of China is the largest supplier of seafood to Japan. The total value of exports slightly decreased in 2012 but rebounded in 2013 (Panel B).
- Total government financial transfers to the Japanese Fisheries sector were more than USD 1.8 billion. In 2012, direct payment decreased by 61%, and cost-reducing transfers increased by 209% compared with average from 2005 to 2010, while total GFT decreased 8.4%. The vast majority of support is for general services (Panel C).
- The ageing of Japanese workers in the fishing sector is an important issue. In 2012, it was reported that the number of fishermen decreased by 21% from the base period (2005-10). However, this figure is affected by the great earthquake. The total number of vessels decreased by 15% compared with base period (Panel D).

Key fisheries indicators

Panel A. Key species by value in 2012

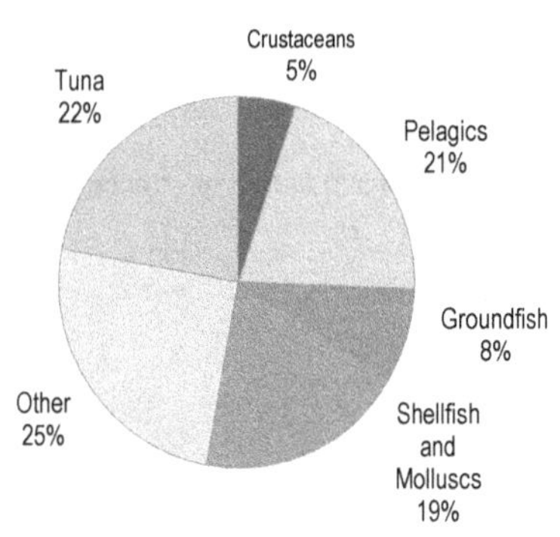

Panel B. Trade evolution

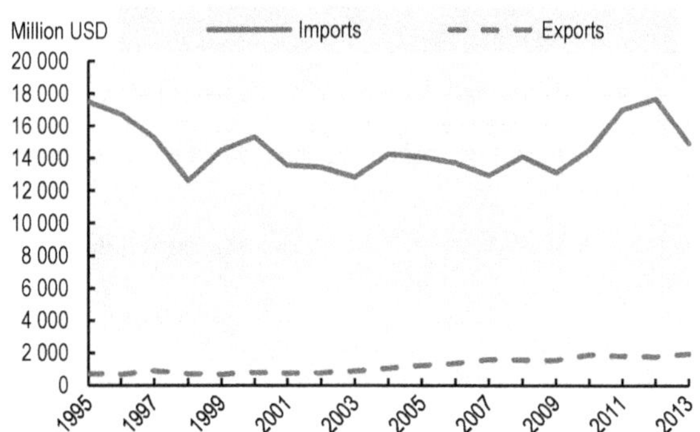

Panel C. Evolution of government financial transfers

USD	Average 2005-10*	2012	% change
Direct payments	14 375 793	5 630 892	-61
Cost reducing	8 423 404	26 012 441	209
General services	1 943 578 360	1 768 449 244	-9

* Or closest available years.

Panel D. Capacity

	Average 2005-10*	2012	% change
Number of fishers	219 818	173 660	-21
Number of fish farmers	..	..	..
Total number of vessels	304 083	258 718	-15
Total tonnage of the fleet	..	..	..

* Or closest available years.

KOREA

Summary and recent developments

- In 2012, the general trends of increasing aquaculture and decreasing coastal capture fisheries continued. Increases in aquaculture were driven by seaweed farming such as laver and kelp. Distant water fisheries was the only sector that recorded strong growth. Trade value and volume rose slightly.
- To ensure the sustainability of the marine ecosystem, the Korea Fisheries Resources Agency (FIRA) was established in 2011. As of 2012, FIRA restored 26 coastal sites to improve spawning grounds and fisheries habitats, particularly by planting sea grass in an area of 1 946 ha. In 2013, 10 May was designated as Marine Gardening Day to raise public awareness of the seriousness of ocean devastation and the importance of restoring marine ecosystem. This day is celebrated by planting marine seaweed and aquatic plants to reinvigorate the health of marine ecosystem.
- To fight against IUU fishing overseas by Korean-flagged distant water fishing vessels as well as Korean Nationals, the Government of Korea amended the Distant Water Fisheries Development Act in 2014 and 2015. The amendments introduced more severe sanctions on IUU fishing as follows: an imprisonment of up to five years; or a criminal fine of up to five times the wholesale value of illegal catches, or a criminal fine of at least KRW 500 million up to KRW 1 billion, whichever is higher; mandatory installation of VMS on all distant water fishing vessels and carrier vessels; pre-authorisation requirements for transshipment; and monitoring of all distant water fishing vessels.
- To improve the quality of fish products, an improved management system compatible to international standards was established, in accordance with the Agricultural and Fishery products Quality Control Act. Since the first pilot projects in 2005, the Fishery Traceability Scheme has expanded to now include more than 20 major species, such as flatfish, oysters, warty sea squirt, abalone, laver, and kelp.

Capture fisheries and aquaculture production

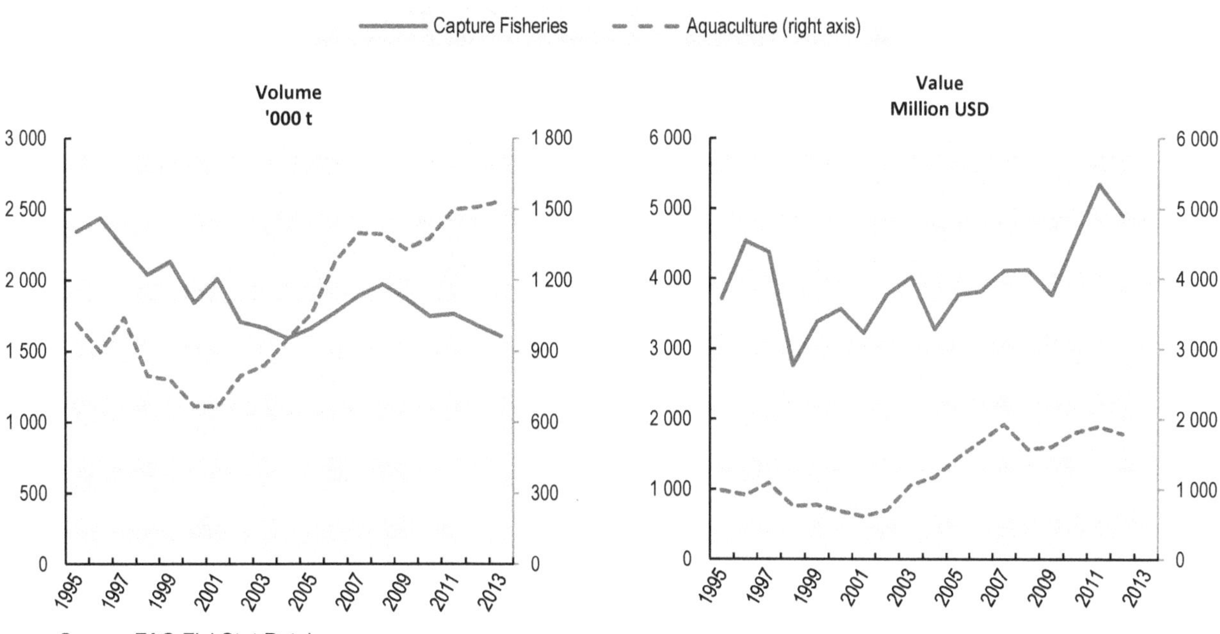

Source: FAO FishStat Database.

Key characteristics of Korean fisheries

- The most important species landed in 2012 in terms of value were pelagics (27%), followed by shellfish and molluscs (24%). The main species in coastal fisheries are anchovies, mackerel, squids, and cutlass fish. Seaweeds such as laver, sea mustard, and kelp, account for around 70% of total aquaculture production (Panel A).
- The trade gap between fish and fishery products in value terms widened in 2012. Major export destinations were Japan, People's Republic of China (hereafter "China"), Thailand, and the United States. Traditional markets such as Japan and China weakened, but this was more than compensated by increases to newer markets such as the United States, and the European Union. Major imported species included shrimp, octopus, Pollack, croaker, and squid, predominantly sourced from China, the Russian Federation, and Viet Nam (Panel B)
- A total of USD 342 million was transferred to the Korea's fisheries sector in 2011, a decrease of 44%, compared to the previous period. Government Financial Transfers to general services dropped by 42% (Panel C).
- The number of fisher and fishing vessels decreased by 17% and 14% respectively between base period and 2013, While the number of fish farmers decreased by 4%, aquaculture production still enjoyed a slight increase, indicating significant improvements in labour productivity (Panel D).

Key fisheries indicators

Panel A. Key landings in value, 2013

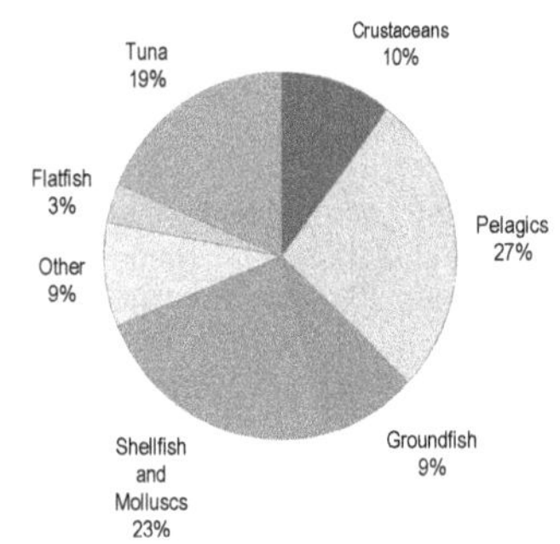

Panel B. Trade evolution

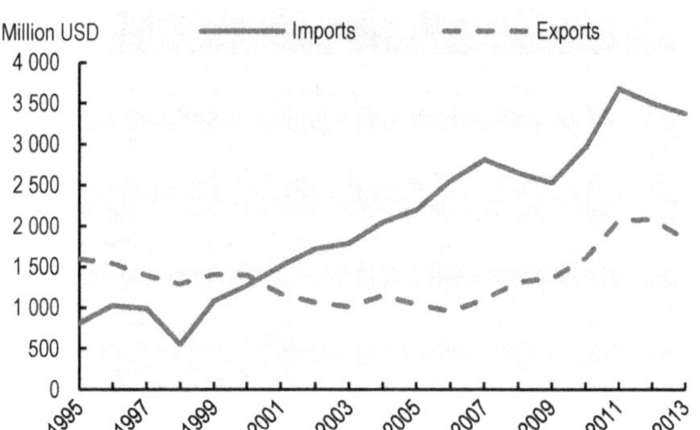

Panel C. Evolution of government financial transfers for marine capture fisheries

USD	Average 2005-10*	2011	% change
Direct payments	124 218 009	33 607 803	-73
Cost reducing	34 182 392	46 571 742	36
General services	453 980 702	261 943 809	-42

* Or closest available years.

Panel D. Capacity

	Average 2005-10*	2013	% change
Number of fishers	82 021	68 337	-17
Number of fish farmers	42 622	40 974	-4
Total number of vessels	82 988	71 287	-14
Total tonnage of the fleet	642 522	607 224	-5

* Or closest available years.

LATVIA

Summary of recent developments

- The Fisheries Law was amended to modify coastal fisheries management by defining all coastal quotas as a percentage of the Latvian total quota for the main regulated species in the Baltic Sea as well as allowing within the year for unused quota to be transferred from coastal to offshore fisheries. Other changes include a ban on gillnets in subsistence fisheries in public inland waters, introduction of some limits on fishing in inland waters, and a measure to combat IUU fishing that introduces new competencies to supervise Latvian nationals on vessels flagged outside Latvia.
- The Latvian Fisheries Integrated Control and Information System was reconstructed to integrate all data from issue of license to data obtained from Fishing Vessel Register, covering whole chain "from net to plate" - fishing activities (logbook information), landings, first sales, etc., as well violations of rules and penalties.
- The situation in the Russian Federation and Ukraine has put pressure especially on exports of canned fish, which are usually marketed in those two countries.
- In order to manage EU financial assistance Operation Programme for the implementation of European Maritime and Fisheries Fund 2014-2020 was elaborated in 2013-2014. Major stress was put on increase of production efficiency and productivity, as well innovations were acknowledged as one of milestones where funding is essential.

Capture fisheries and aquaculture production (tonnes)

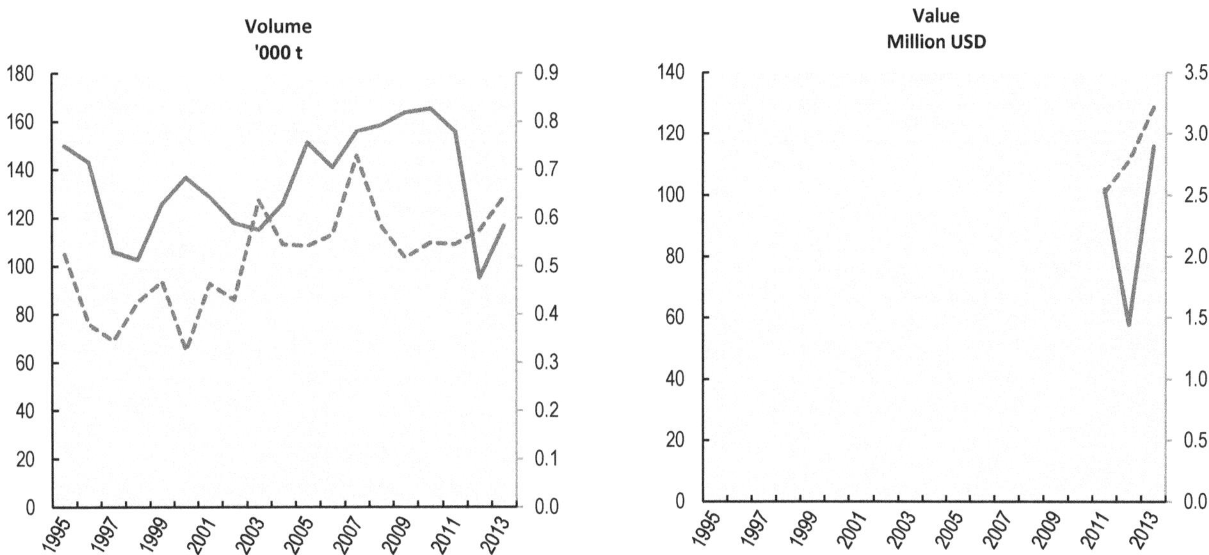

Source: FAO FishStat Database.

Key characteristics of Latvian fisheries

- Pelagic species (Herring and Sprat) are the mainstay of Latvian harvests, accounting for almost 90% of the total (Panel A).
- Latvia has made considerable efforts to expand trade opportunities, and the sector has become much more exposed to global trade (Panel B)
- The majority of financial transfers are made as part of the European Fisheries Fund (EFF) (Panel C).
- Employment has been stable, with the majority of fishers participating in the small coastal fishery (Panel D).

Key fisheries indicators

Panel A. Key landings in value, 2013

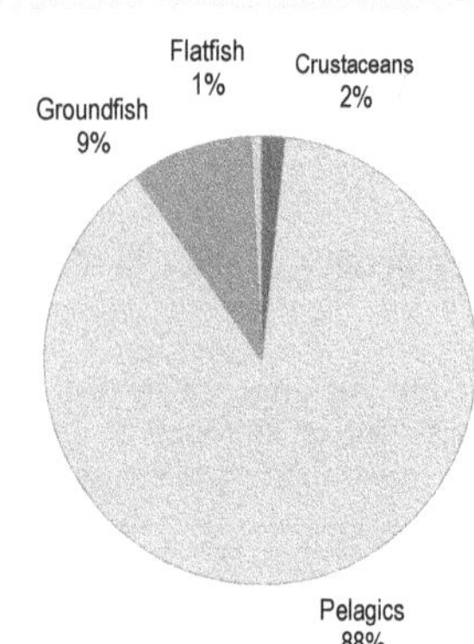

Panel B. Trade evolution

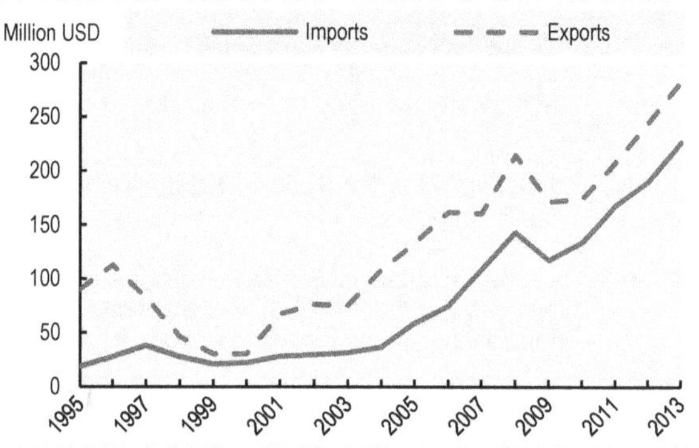

Panel C. Evolution of government financial transfers for marine capture fisheries

USD	Average 2005-10*	2013	% change
Direct payments	..	3 100 550	..
Cost reducing	..	..	..
General services	..	..	..

* Or closest available years.

Panel D. Capacity

	Average 2005-10*	2013	% change
Number of fishers	..	701	..
Number of fish farmers	..	355	..
Total number of vessels	..	703	..
Total tonnage of the fleet	..	29 945	..

* Or closest available years.

MEXICO

Summary and recent developments

- Mexican fisheries encompass a number of activities in the Pacific and Atlantic Oceans, and its extensive brackish and fresh water bodies. It includes capture, culture, transformation and commercialisation activities. Fisheries are very important to the national economy and a vital source of nutrition for Mexicans. It is also an important source of foreign currency. At the community level, fisheries activities provide essential income for certain parts of the population and are a driving force for regional economic development.
- Mexico has around 11 592 km of shoreline with 3 million square km of Exclusive Economic Zone (EEZ) over 2.9 million hectares of inland waters, including 1.6 million of coastal lagoons. It also has a privileged location with the influence of important marine currents that provide a large biodiversity in marine, brackish and fresh waters.
- The fisheries and aquaculture sector is governed by five strategic areas: Fisheries and Aquaculture Management integrally and sustainably, Compliance and Enforcement Policy, Strategic Development of Aquaculture, Fostering the Capitalisation of Fisheries and Aquaculture, and Promoting the consumption of fishery and aquaculture products.
- The Mexican Government has a long term vision to encourage the national development, competitiveness and the strategic planning for fisheries and which is articulated through the National Development Plan, the Sectorial Program of Farming and Fishing of the SAGARPA, and the National Sector Program of Fisheries and Aquaculture. There is a particular focus on improving the competitiveness of the fisheries sector. This could be achieved through natural resource sustainability, updating the regulatory framework to protect the sector's interests abroad, the integration of productive chains, the continued support for innovative projects throughout the country, and encouraging regional development by promoting small-scale projects within the rural sector.

Capture fisheries and aquaculture production

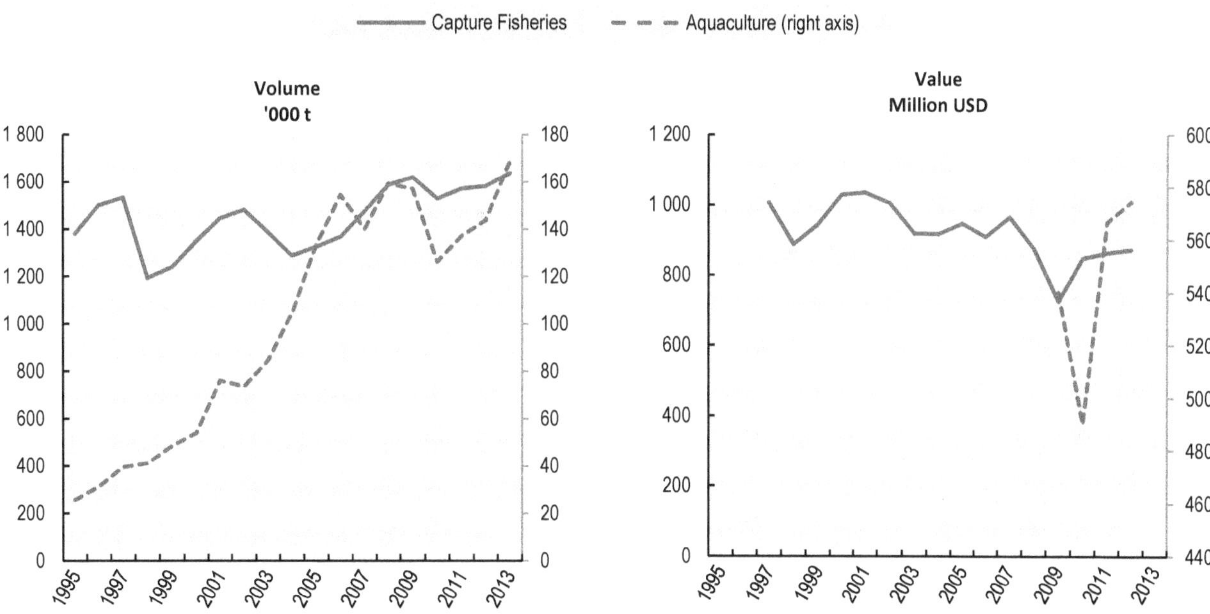

Source: FAO FishStat Database.

Key characteristics of Mexican fisheries

- In 2013, production volume was 1 746 000 tonnes, up 3.48% from 2012, continuing the upward trend since 2005. National production in recent years has remained constant with a slight upward trend mainly because of growth in aquaculture production. Major species include shrimp, clams, crab and tuna. (Panel A)
- In 2013, export volume was 284 495 tonnes worth MXN 14.16 billion (USD 1.1 billion), while imports were 268 151 tonnes worth MXN 12.3 billion (USD 963 million). The main destinations of Mexican fish products exports included the United States, Hong Kong and Spain. Exports included species such as shrimp, tuna, lobster, octopus and sardine. (Panel B)
- A total of MXN 356 million (USD 29.7 million) was transferred to the fisheries sector in 2011. This is a decline of 37% compared to the average between 2005 and 2010. Cost-reducing transfers, the largest category of support, fell by 70% while general services increased by one third. (Panel C)
- Considerable adjustment has taken place recently with a 16% decrease in the number of fishers and a decrease of 6% in the number of vessels between 2012 and the 2005-10 period. Aquaculture has grown significantly in that same period, with 26 000 more employees employed in the sector between these periods. (Panel D)

Key fisheries indicators

Panel A. Key species by value in 2012

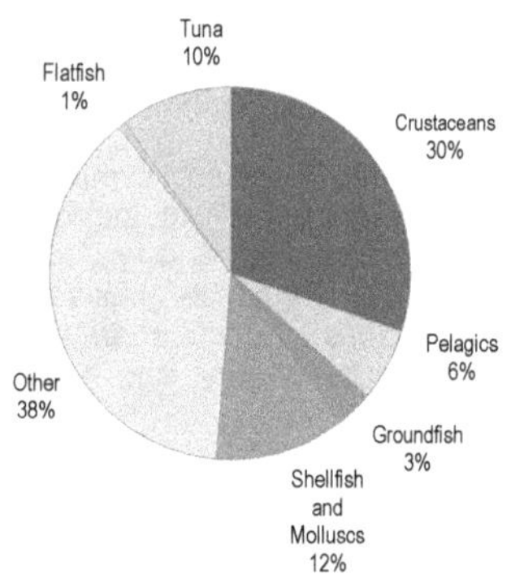

Panel B. Trade evolution

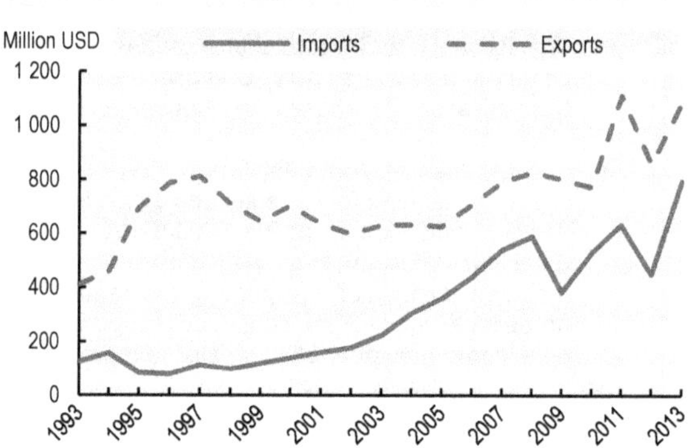

Panel C. Evolution of government financial transfers

USD	Average 2005-10*	2013	% change
Direct payments	6 731 590	3 450 290	-49
Cost reducing	64 031 073	19 260 502	-70
General services	5 204 994	7 013 177	35

* Or closest available years.

Panel D. Capacity

	Average 2005-10*	2013	% change
Number of fishers	249 984	210 247	-16
Number of fish farmers	30 513	56 133	84
Total number of vessels	3 367	3 158	-6
Total tonnage of the fleet	223 131	214 125	-4

* Or closest available years.

NETHERLANDS

Summary and recent developments

- In 2013 the responsibility for managing capture fisheries and aquaculture switched from the Ministry of Economic Affairs, Agriculture and Innovation to the Ministry of Economic Affairs.
- A new type of fishing gear for bottom fishing was introduced in 2005. It uses weak electric pulses to scare flatfish from or out of the seafloor, so the fishing net no longer has to be pulled through the sand to catch the flatfish. This reduces not only the negative impacts of fishing on the surface of the seafloor and thereby its impact on biodiversity, but also reduces wear of the nets and the use of fuel. In 2012-13 the use and development of this fishing method continued. From 2014 onwards the effects of the use of this fishing method on a larger scale will be tested and monitored, in view of the start of the landing obligation for demersal fisheries in 2016.
- In 2012-13 the collection of mussel seed (spat) has been made more sustainable by increasing the use of installations (hanging nets, also known as mussel seed capture installations) for collecting the one to two centimetre small mussels, instead of scraping these young mussels from the tidal flats of the seafloor by fishing vessels. The scraping of mussels is progressively restricted to protect biodiversity and ecosystem.

Capture fisheries and aquaculture production

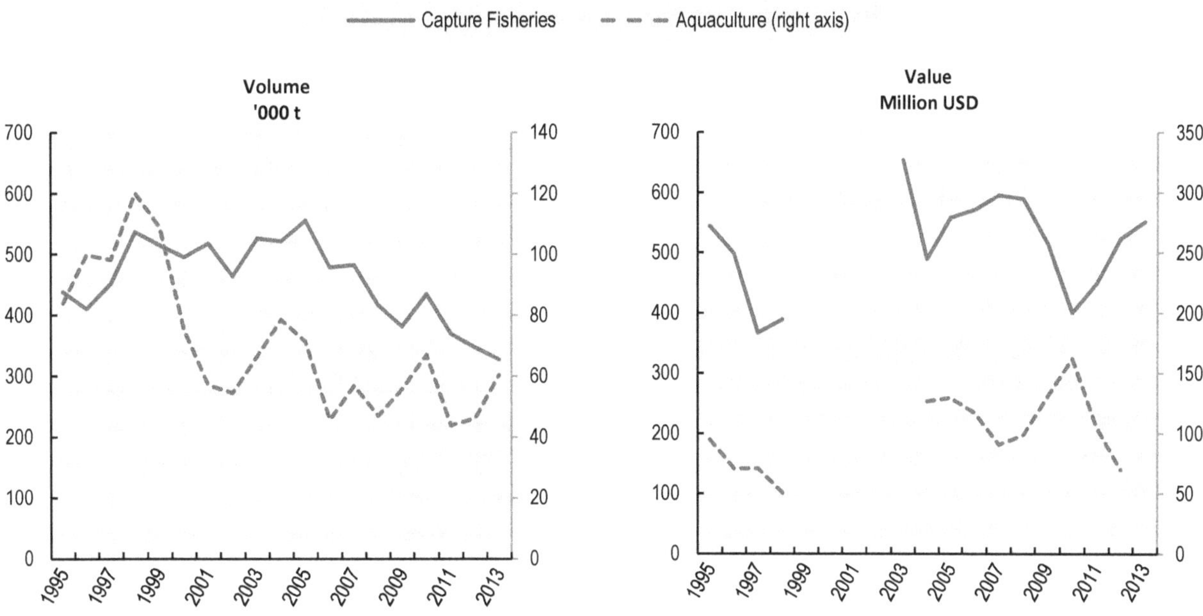

Source: FAO FishStat Database.

Key characteristics of Netherlands fisheries

- Flatfish is by far the most important species harvested by the Dutch fleet and includes in particular sole and plaice. Herring, mackerel and horse mackerel are important pelagic species in the Dutch fisheries (Panel A).
- In 2013 imports decreased (-3%) for a total value of EUR 2 160 million In 2013 imports came mainly from Germany, Belgium, Denmark, United Kingdom, Norway and People's Republic of China. In 2013 the value of exports remained stable at about EUR 2 590 million. In 2013 export was mainly to Belgium (19%), Germany (18%), France (11%) and Italy (10%). Of all the EU-countries, only The Netherlands and Denmark are net exporters (Panel B).
- The principal financial instrument for fisheries is the European Fisheries Fund (EFF). For measures in favour of capture fisheries the Netherlands expended in 2012-13 a total of EUR 20,4 million while EUR 1,7 million was for aquaculture. There were no specific transfers to the fish processing sector. The Dutch operational Programme focuses mainly on innovation and sustainability of the harvesting and aquaculture production (Panel C).
- In 2012 the fleet included around 550 vessels: 14 were active in pelagic fisheries, about 275 were active in North Sea fisheries and 255 in coastal fisheries. The registered inland fleet amounted 70 vessels, but the number of active vessels is unknown. Since 2005 there has hardly been renewal of the fleet and about 90% of vessels are 10+ years while 65% of the vessels are 20+ years (Panel D).

Key fisheries indicators

Panel A. Key landings in value, 2013

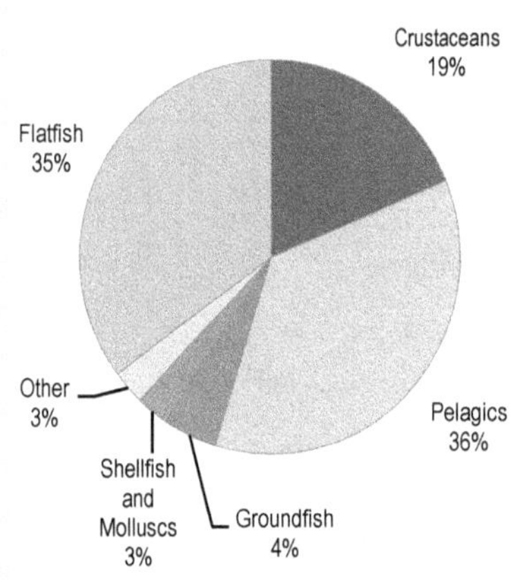

Panel B. Trade evolution

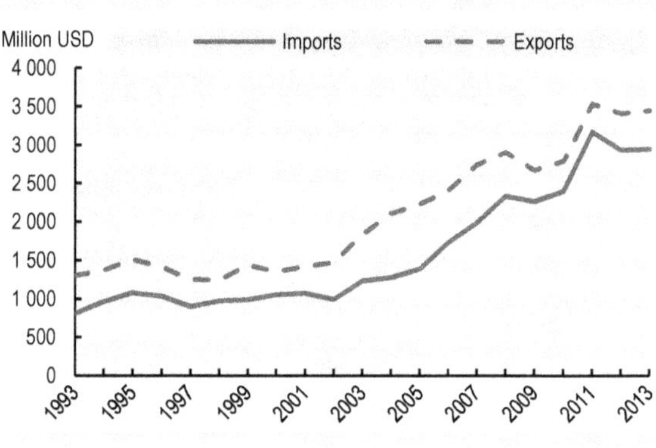

Panel C. Evolution of government financial transfers for marine capture fisheries

USD	Average 2005-10*	2012	% change
Direct payments	13 003 478	..	..
Cost reducing	..	..	..
General services	2 776 685	..	..

* Or closest available years.

Panel D. Capacity

	Average 2005-10*	2013	% change
Number of fishers	2 279	1 975	-13
Number of fish farmers	255	280	10
Total number of vessels	898	904	1
Total tonnage of the fleet	157 446	148 934	-5

* Or closest available years.

NEW ZEALAND

Summary and recent developments

- New Zealand focuses on exports of fisheries products mainly from deep water fishing. The marine harvest is around 600 000 tonnes in volume and approximately NZD 1.5 billion (USD 1.25 billion) per annum in value. The aquaculture industry contributes about NZD 300 million (USD 250 million) per annum. The fisheries industry is the country's fourth or fifth largest exporting sector.
- Using Quota Management System (QMS) based on Individual Transferable Quotas (ITQs), the stock of hoki and orange roughy recovered successfully to a level well within or above the management target range as of 2013.
- Under the "Fisheries 2030" which is a Cabinet-endorsed strategy for fisheries, an objective-based approach to fisheries management in being developed through establishing "National Fisheries Plans". New Zealand participates in international marine protection measures including the revised National Plan of Action for the Conservation and Management of Sharks and has ratified the FAO Agreement on Port State Measures to Prevent, Deter, and Eliminate IUU Fishing.
- In 2011, a package of aquaculture reforms came into effect designed to aid the development of the sector. An Aquaculture Strategy and a Five-year Action Plan were adopted in 2012.

Capture fisheries and aquaculture production

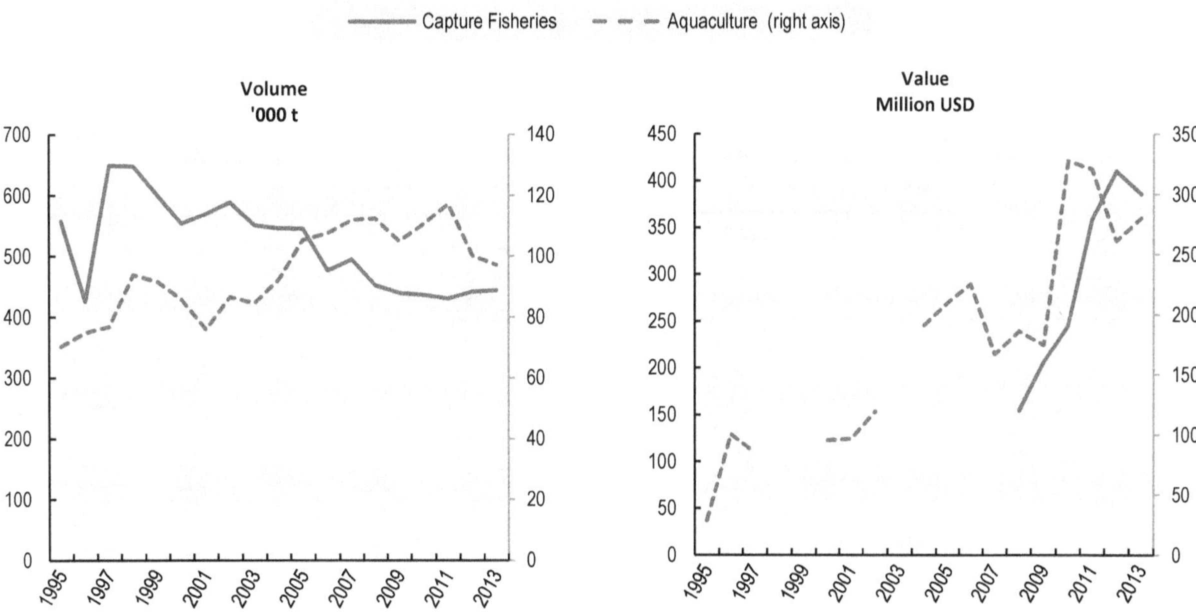

Source: FAO FishStat Database.

Key characteristics of New Zealand fisheries

- Crustaceans and Groundfish account for 60% of landings by value in 2012. Top species in capture fisheries landed by weight in fishing year 2012-13 were hoki, squid, and mackerel. Aquaculture is primarily based on three species: green lipped mussels, king salmon, and pacific oysters (Panel A).
- Approximately 90% of New Zealand's seafood production is exported. Recently, exports volume stays stable after rapid growth from 2009. The largest importer for New Zealand's fisheries products is People's Republic of China (27%), Australia (18%), the European Union (11.5%), the United States (10.7%), followed by Japan and Hong Kong, China (Panel B).
- Most of the Government Financial Transfers were general services. The amount transferred to fisheries sector has been increasing to USD 68 million, i.e. by 12%. There is cost recovery charges associated with fisheries management service and conservation services (Panel C).
- In 2013, the number of fishermen in the marine fisheries sector increased by 11%, up from previous base period. However, total number of vessels and tonnage of the fleet decreased (Panel D).

Key fisheries indicators

Panel A. Key landings in value, 2012

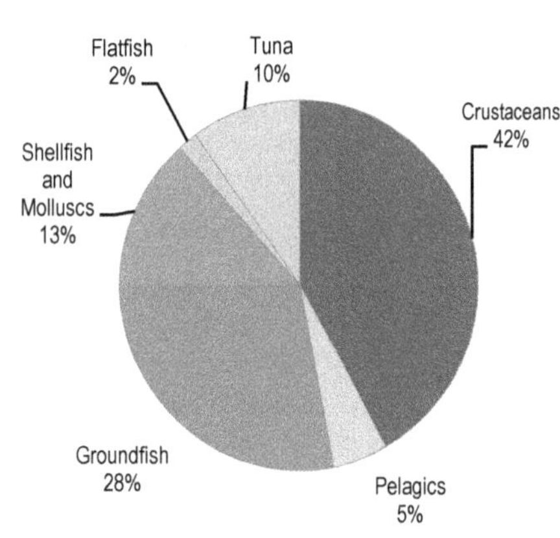

Panel B. Trade evolution

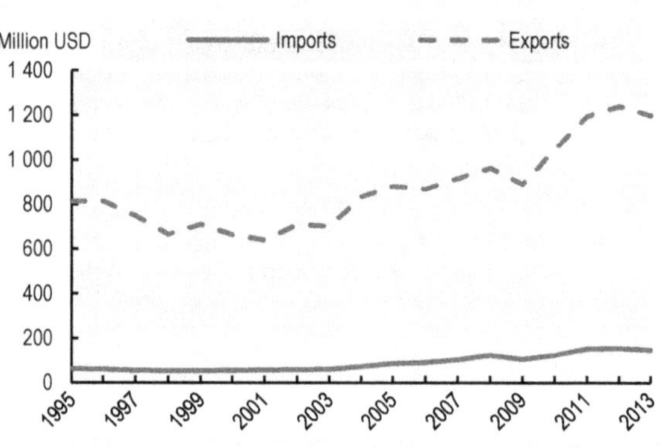

Panel C. Evolution of government financial transfers for marine capture fisheries

USD	Average 2005-10*	2013	% change
Direct payments	..	..	..
Cost reducing	..	..	..
General services	60 953 411	68 254 935	12

* Or closest available years.

Panel D. Capacity

	Average 2005-10*	2013	% change
Number of fishers	1 580	1 746	11
Number of fish farmers	686	760	11
Total number of vessels	1 497	1 367	-9
Total tonnage of the fleet	140 123	122 111	-13

* Or closest available years.

NORWAY

Summary and recent developments

- In 2013, landings of fish by Norwegian registered vessels totalled 2.1 million tonnes, with a total first-hand value of NOK 12.5 billion. This implies a decrease in catch from the 2012-level of 60 000 tonnes, and a decrease in value of NOK 1.7 billion.
- Aquaculture production of Atlantic salmon and rainbow trout decreased from 1 306 772 tonnes in 2012 to 1 238 554 tonnes in 2013. The first-hand value of these farmed species amounted to NOK 29.6 billion in 2012 and NOK 39.8 billion in 2013. Farming of other marine species is modest.
- The overall value of Norwegian seafood exports in 2013 was NOK 61 billion, a substantial increase from NOK 52.1 billion in 2012. Of this, exports of capture fish accounted for NOK 18.8 billion whereas farmed fish accounted for NOK 42.2 billion.
- The state of the most important commercial fish stocks in Norwegian fisheries is good. The structural quota system (SQS), which was endorsed by Parliament in 2007, continues with small adjustments.
- A new Act concerning first-hand sales of wild living marine resources (fish sales organisations act) was adopted in June 2013. This law replaces the Raw Fish Act from 1951.
- Norway has continued efforts to combat IUU-fishing and organised crime related to fisheries at the national and international levels. Norway has prioritised efforts on issues related to discards and by-catch of marine recourses and the protection of vulnerable marine ecosystems (VMEs).

Capture fisheries and aquaculture production

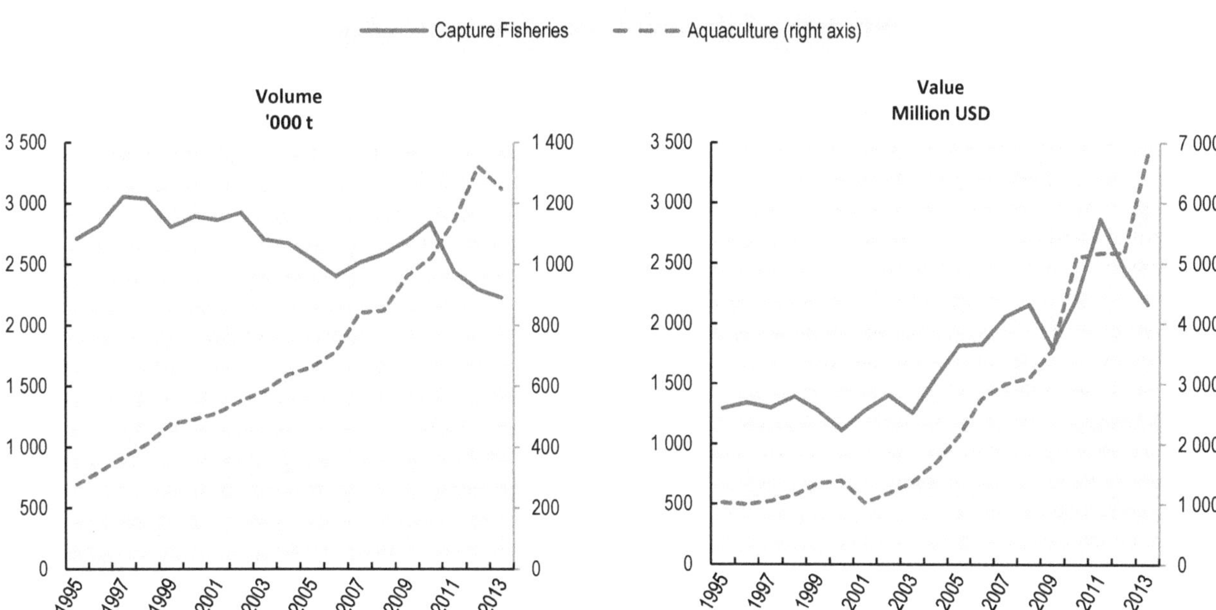

Source: FAO FishStat Database.

Key characteristics of Norwegian fisheries

- Preliminary figures indicate that the total Norwegian landings decreased from about 2.14 million metric tonnes in 2012 to 2.08 million tonnes in 2013. The total first-hand value decreased from NOK 14.2 billion in 2012 to NOK 12.5 billion in 2013. The main species harvested in 2013 were herring (506 230 tonnes), cod (469 932 tonnes) and mackerel (164 608 tonnes) (Panel A).
- The overall value of Norwegian seafood exports in 2013 was NOK 61 billion compared to NOK 52.1 billion in 2012. The most important export market for Norway was the European Union, taking 59% of all seafood exports from Norway. Approximately 95% of the Norwegian production of seafood is exported and Norway is the second largest seafood exporter globally (Panel B).
- A total of NKR 1,35 billion was expended by Norway on general services in 2013 and 11% increase in the previous year. Major expenditures are for surveillance and enforcement and for research. Minor support is available to the sealing sector and for the transport scheme (of fish) established to reduce cost disadvantages caused by geographical or structural conditions. NKR 3,45 million were expended on social measures (unemployment scheme) in 2013.
- The total number of commercial fishermen in Norway was 11 602 in 2013, a decrease from 12 048 in 2012. The number of fishing vessels registered in the *Register of Norwegian Fishing Vessels* decreased from 6 211 vessels in 2012 to 6 128 vessels in 2013. The largest vessel group consists of the smallest coastal vessels (less than 11 meters) and account for 79% of the vessels. Approximately 5 700 persons were registered as directly employed in the aquaculture sector in 2013. When including spin-off effects, studies show that the industry contributed approximately 23 700 man-years in 2012

Key fisheries indicators

Panel A. Key landings in value, 2013

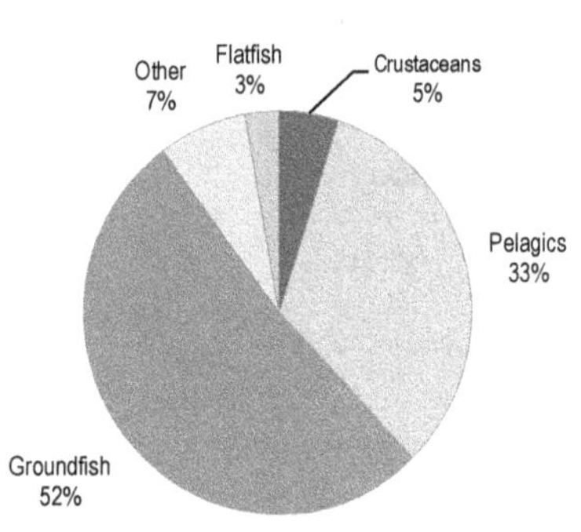

Panel B. Trade evolution

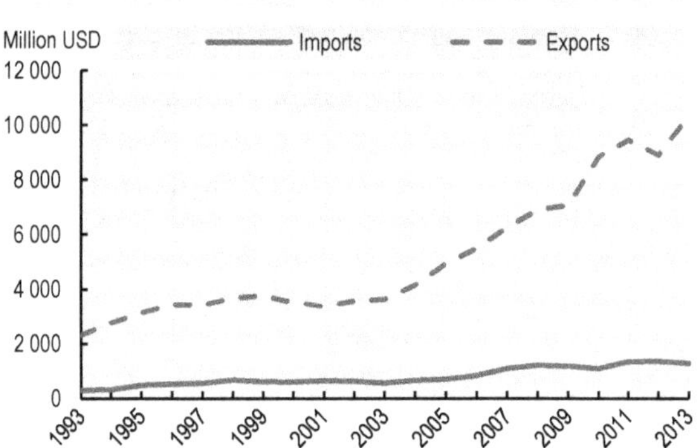

Panel C. Evolution of government financial transfers for marine capture fisheries

USD	Average 2005-10*	2013	% change
Direct payments	3 184 129	8 168 127	157
Cost reducing	58 735 836	53 710 015	-9
General services	196 437 677	274 181 230	40

* Or closest available years.

Panel D. Capacity

	Average 2005-10*	2013	% change
Number of fishers	14 009	11 601	-17
Number of fish farmers	4 860	5 950	22
Total number of vessels	6 944	6 126	-12
Total tonnage of the fleet	364 368	381 170	5

* Or closest available years.

POLAND

Summary and recent developments

- Harvest, consumption, imports and exports all increased between 2012 and 2013, both in quantity and value terms. This indicates that a broad-based expansion is taking place in the fishing sector.
- The number of fishing vessels and employment is also increasing in the fishing sector. After a period of privatization, the fishing sector has become one of the most rapidly developing branches of the food processing sector.

Capture fisheries and aquaculture production

Capture Fisheries — Aquaculture (right axis)

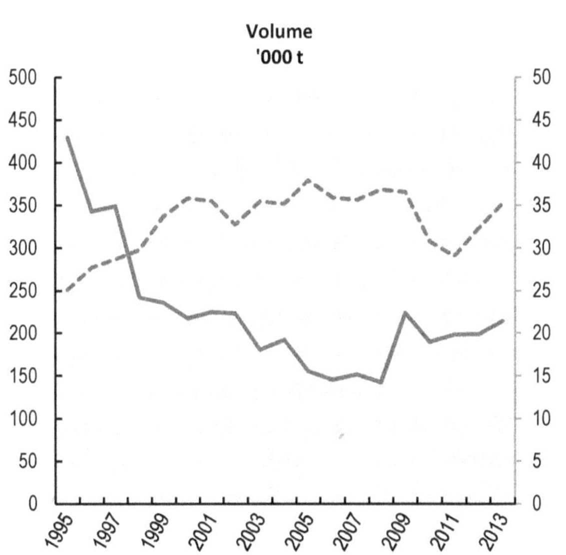

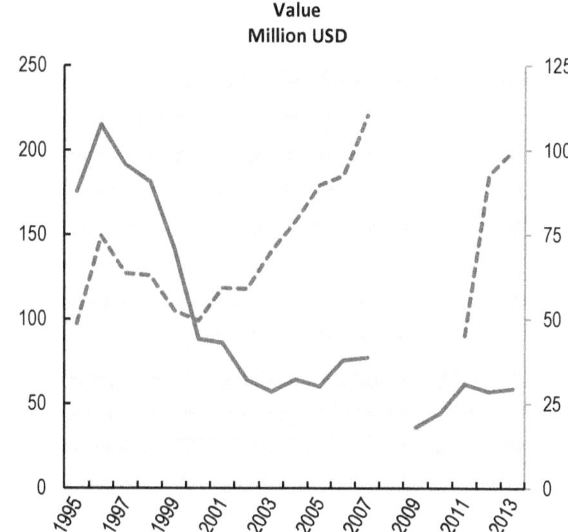

Source: FAO FishStat Database.

Key characteristics of Polish fisheries

- Pelagics such as sprat and horse mackerel are the most important domestic species harvested. The Baltic sea and Atlantic Ocean are the main fishing areas (Panel A).
- Imports increased significantly by value in 2013 reflecting higher import prices. Exports are mainly to EU countries. Imports are dominated by raw materials while exports are mainly processed products, of which Salmon is the most important (Panel B).
- Government financial transfers are made under the EU Operational Programme "Sustainable Development of the Fisheries Sector and Coastal Fishing Areas 2007-2013 (Panel C).
- It is estimated that 5300 workers are employed in the fish farming sector, while fishing and fish processing employ a total of 28 500 (Panel D).

Key fisheries indicators

Panel A. Key landings in value, 2013

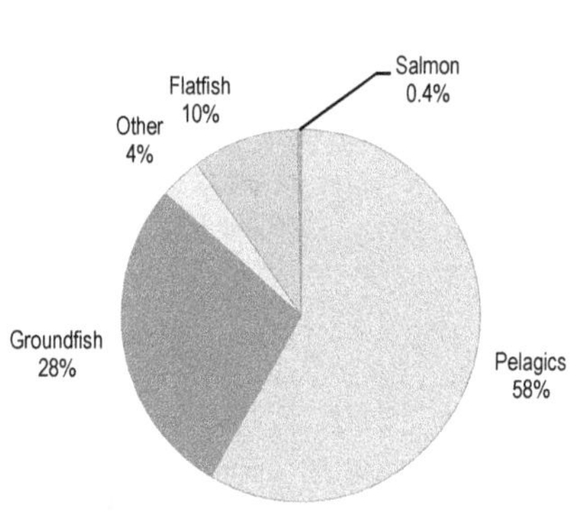

Panel B. Trade evolution

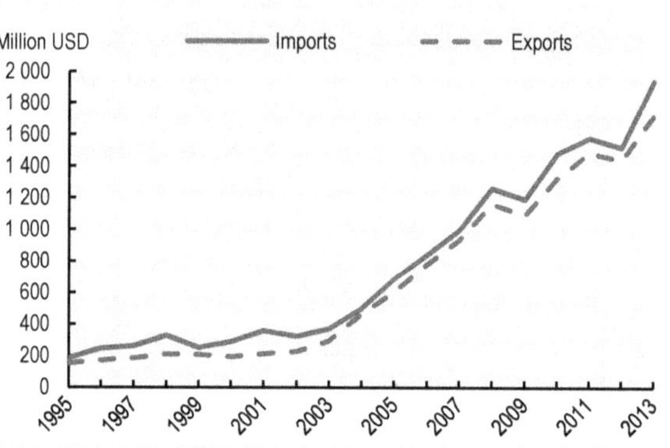

Panel C. Evolution of government financial transfers

USD	Average 2005-10*	2011	% change
Direct payments	33 668 110	10 789 685	-68
Cost reducing	310 285	..	..
General services	11 379 315	..	..

* Or closest available years.

Panel D. Capacity

	Average 2005-10*	2013	% change
Number of fishers	3 899	..	..
Number of fish farmers	3 640	5 430	49
Total number of vessels	860	899	5
Total tonnage of the fleet	34 965	33 949	-3

* Or closest available years.

PORTUGAL

Summary and recent developments

- The Ministry for Agriculture, the Sea, the Environment and Spatial Planning (MAMAOT) was replaced with the Ministry for Agriculture and the Sea (MAM). A General Direction for Natural Resources, Security and Maritime Services, a General Direction for Sea Policies, a Portuguese Agency for the Environment and an Institute for the Conservation of Nature and Forests were also created.
- In 2012 and 2013, new regulations have been adopted and implemented for sardine fisheries (Regulation No. 251/2010), which restrict harvest, sale, fishing periods, and landing sites.
- A new simplified licensing system has been set up for offshore aquaculture production sites.
- Portugal remains the EU country with the highest per capita fish consumption (54 Kg per capita/year).
- In 2012-13, almost 600 projects were approved under the programme PROMAR, together accounting for eligible investment of about EUR 120 million, of which 60 will be financed by the European Union and 20 by the national budget.

Capture fisheries and aquaculture production

Capture Fisheries — Aquaculture (right axis)

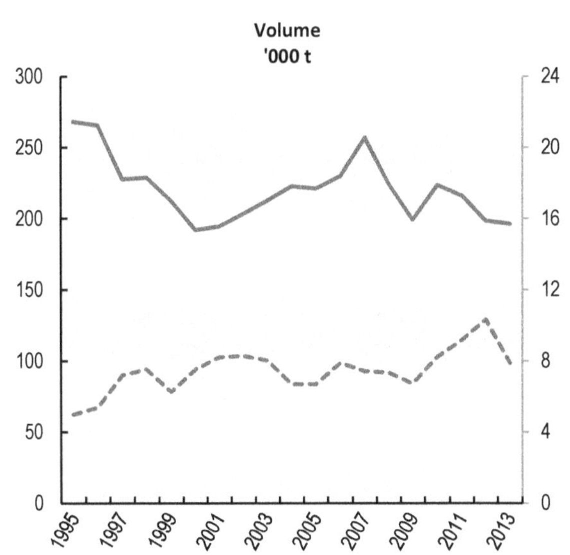

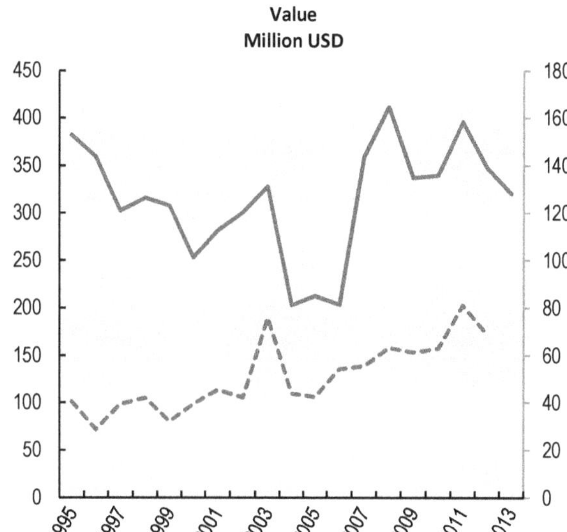

Source: FAO FishStat Database.

Key characteristics of Portuguese fisheries

- In 2013, the Portuguese fleet harvest about 200 000 tonnes of fish, slightly down from 2012. The drop notably results from lower volumes of sardine due to new regulations. Volumes harvested in external waters slightly increased.
- The total value of the catch also went down, notably as a result of decreasing prices for octopus, mackerel and jack mackerel.
- In December 2013, the Portuguese fleet counted over 8000 vessels, with a total capacity of almost 100 000 GT and over 365 000 KW. The number of vessels decreased over the last decade but capacity remained stable.
- In 2012, total aquaculture production reached over 10 000 tonnes, for a value of almost EUR 54 million. This is up 12% in volume compared to 2011, but down 8% in value, due to lower mollusc prices.
- In 2013, the trade balance deficit was over EUR 640 million (USD 850 million), slightly up since 2012, due to increasing exports.

Key fisheries indicators

Panel A. Key landings in value, 2013

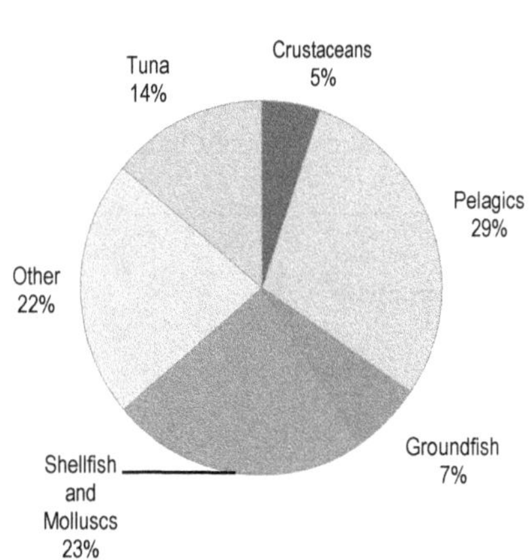

Panel B. Trade evolution

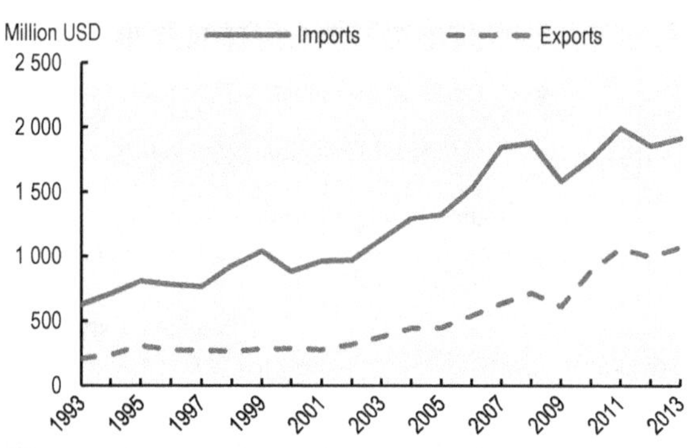

Panel C. Evolution of government financial transfers for marine capture fisheries

USD	Average 2005-10*	2013	% change
Direct payments	14 302 778	12 845 240	-10
Cost reducing	349 983	580 822	66
General services	21 993 344	16 467 263	-25

* Or closest available years.

Panel D. Capacity

	Average 2005-10*	2013	% change
Number of fishers	17 259	16 797	-3
Number of fish farmers	..	..	..
Total number of vessels	8 823	8 232	-7
Total tonnage of the fleet	105 775	99 917	-6

* Or closest available years.

SLOVAK REPUBLIC

Summary and recent developments

- The fisheries sector in the Slovak Republic consists of aquaculture and fish processing. There is little commercial marine and inland capture fishery.
- Total aquaculture production in 2013 was 1 084.8 tonnes. This is less than 2012 production of 1 287.1 tonnes but an increase over 2011 (814.7 tons).
- Recreational catches average around 1 900 tonnes annually, making angling more important than aquaculture in terms of volume.

Aquaculture production

Volume

'000 t

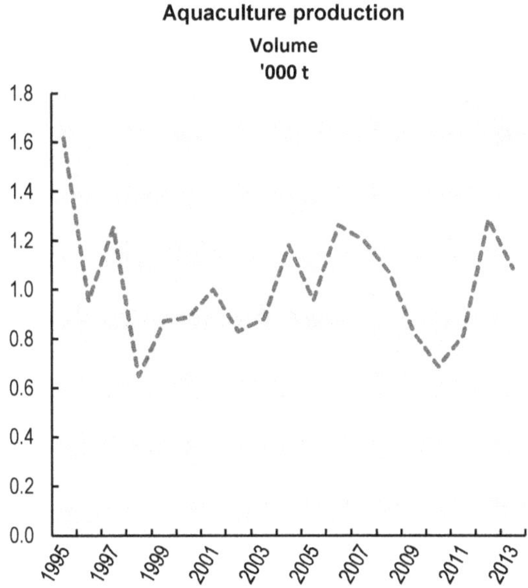

Source: FAO FishStat Database.

Key characteristics of Slovak Republic fisheries

- Slovakia has no commercial capture fisheries. The composition of key species from aquaculture in terms of volume is stable over the long term. Rainbow trout represents about 76% of the aquaculture production while carp is around 20%.
- GFTs are provided to the aquaculture sector as well as to the marketing and processing sector. The GFTs are funded by the European Fisheries Fund 2007 - 2013.
- The number of full-time employees in fish farms increased in 2013, while the number of seasonal employees declined. Overall the workforce is stable, indicating that the full time workforce increased due to a change in state policy regarding employment (Panel C).

Key fisheries indicators

Panel A. Key aquaculture species in value, 2009

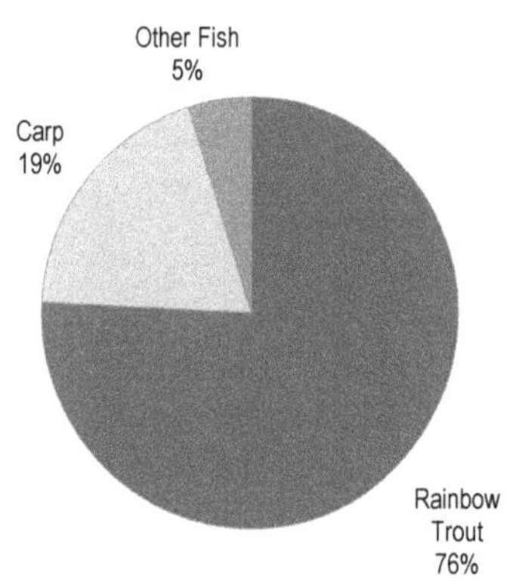

Panel B. Trade evolution

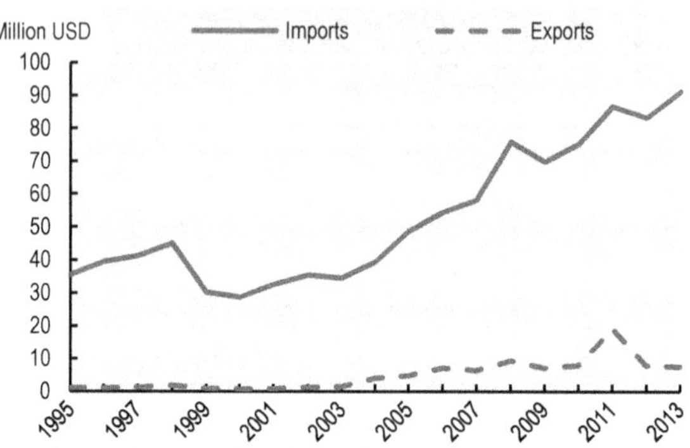

Panel C. Capacity

	Average 2005-10*	2011	% change
Number of fishers	..	..	..
Number of fish farmers	784	929	18
Total number of vessels	..	..	..
Total tonnage of the fleet	..	..	..

* Or closest available years.

SPAIN

Summary and recent developments

- The central government has sole jurisdiction over sea fishing and therefore establishes legislation on sea fishing in exterior waters. "Basic legislation" on management of fisheries and on management of the commercial activity is also established by the central government; this implies that Autonomous Communities can adopt provisions that complement legislation in these two areas. They also have sole jurisdiction over fishing in internal waters, the harvesting of shellfish, and aquaculture.
- The Ministry carries on initiatives under the Marine Nature Network. After the creation of the Marine Protected Area (MPA) "El Cachucho" in 2008 with, the Project LIFE+ INDEMARES 2009-2013 has prepared grounds for protecting particular offshore areas such as submarine canyons and mountains, methane volcanoes, pockmarks and key areas for cetaceans, sea turtles and sea birds. In 2014, new MPAs will be defined based on the information gathered through the project: 10 Sites of Community Importance (SCI) and 39 Special Bird Areas (SBA) will be created in the framework of the Natura 2000 Network in the sea.
- Fight against illegal, unreported and unregulated, IUU, fishing, has become a priority of the Spanish administration. Existing regulations are being reformed (including the National Act) while new initiatives are being undertaken at international level.
- Two new Regulations were adopted in 2013. The first one creates a national registry of serious infringements to the EU Common Fisheries Policy (CFP) and updates the amount of sanctions and creates rules for taking back licenses. The second Regulation defines the Spanish Economic Exclusive Zone in the Northwest Mediterranean sea.

Capture fisheries and aquaculture production

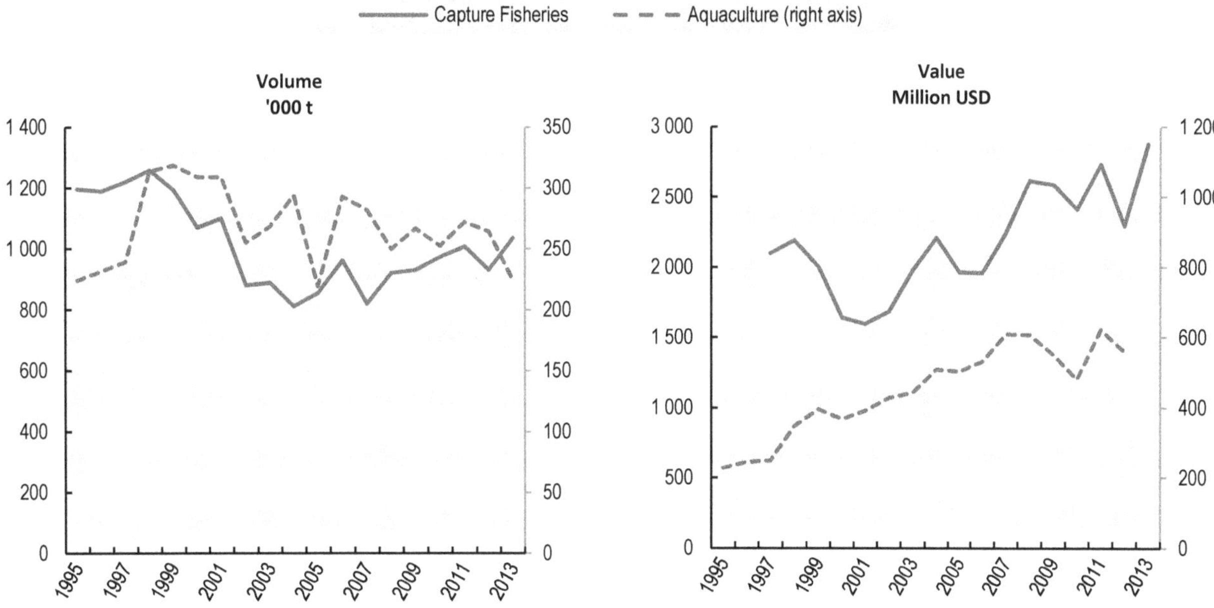

Source: FAO FishStat Database.

Key characteristics of Spanish fisheries

- In 2012, the fishery and aquaculture sector produced about 813 000 tonnes of fishery products, of which about one quarter from aquaculture. Capture production was valued at about EUR 1.8 billion (USD 2.3 billion), and aquaculture production around EUR 435 million (USD 578 million, that is about 20% of the total value generated). At the end of 2012, the sector employed a total of over 55 000 people of which about 35 000 in the capture sector and 20 000 in aquaculture.
- In 2013, Spanish imports of fisheries products amounted to 1.5 million tonnes, with a value of almost EUR 5 billion (USD 6.6 billion). Exports amounted to 1 million tonnes and almost EUR 3 billion (around USD 4 billion). About 70% of imports are supplied by third countries; the main suppliers being the People's Republic of China, Argentina, Morocco and Ecuador. 74% of Spanish exports of fisheries products go to the EU market.
- Total support awarded to the sector amounted to EUR 155 million (USD 206 million) in 2012, and EUR 86 million (USD 114 billion) in 2013, of which, respectively, EUR 106 million (USD 141 million) and EUR 59 million (USD 78 million) were funded via the European Fisheries Fund and the remaining by Spain.
- As of 31 December 2013, the Spanish fishing fleet counted over 9 331 motor vessels and 540 small vessels. In 2012 and 2013 support for permanent cessation was awarded to a total of 276 fishing vessels, reducing the overall tonnage of the fleet by 20 000 GT.

Key fisheries indicators

Panel A. Key species by value in 2013

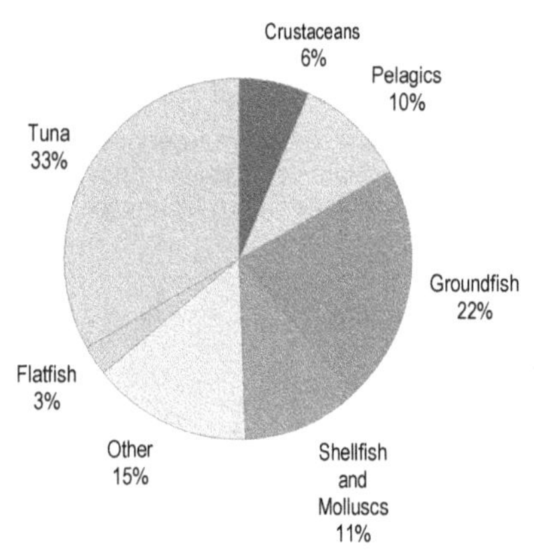

Panel B. Trade evolution

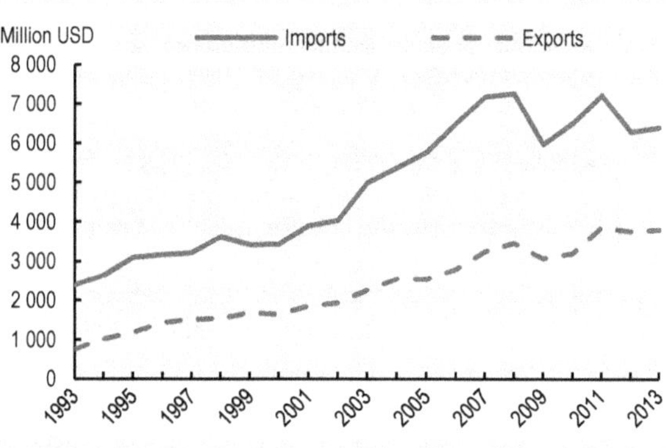

Panel C. Evolution of government financial transfers

USD	Average 2005-10*	2013	% change
Direct payments	77 113 259	64 656 849	-16
Cost reducing	72 008 090	..	..
General services	53 596 516	20 085 756	-63

* Or closest available years.

Panel D. Capacity

	Average 2005-10*	2013	% change
Number of fishers	36 086	35 671	-1
Number of fish farmers	6 716	19 892	196
Total number of vessels	12 242	9 872	-19
Total tonnage of the fleet	457 895	372 617	-19

* Or closest available years.

SWEDEN

Summary and recent developments

- The general principles governing national fisheries policy are established in a Parliament Act. The Parliament has recently decided to change the Parliament Act so that legal persons, and not only natural persons, can be granted a fishing licence. Natural persons do not need to have personal fishing licence anymore, instead it will be sufficient to have a fishing licence for the fishing boat.
- In 2012, Sweden allowed vessels with cod permits in the Baltic Sea to transfer days between them to facilitate an efficient management of fishing opportunities.
- In 2013, the Swedish Agency for Marine and Water Management decided on new regulations for the Swedish fishing of salmon in the Baltic Sea. Salmon fishing with longlines in the Baltic Sea was prohibited and the Swedish quota was allocated solely to trap fisheries.
- Awaiting implementation at EU level, Sweden introduced in national legislation technical measures which has resulted in increased selectivity for demersal and Pandalus fisheries in the Skagerrak.
- In 2012, the government assigned the Parliamentary Committee on Environmental Objectives to develop a Swedish strategy for a coherent and sustainable water policy, with intermediate targets and policy instruments for achieving these goals. The strategy was presented to the government in late June 2014.
- In January 2014, the new EU Common Fisheries Policy (Basic Regulation (Reg (EU) No. 1380/2013) and Common Market Organisation (Reg (EU) No. 1379/2013)) entered into force. Efforts in the coming years will focus on successfully implementing the new policy, at the national level as well as within the European Union in the framework of regionalisation in the Baltic and the North Sea areas.

Capture fisheries and aquaculture production

Capture Fisheries — Aquaculture (right axis)

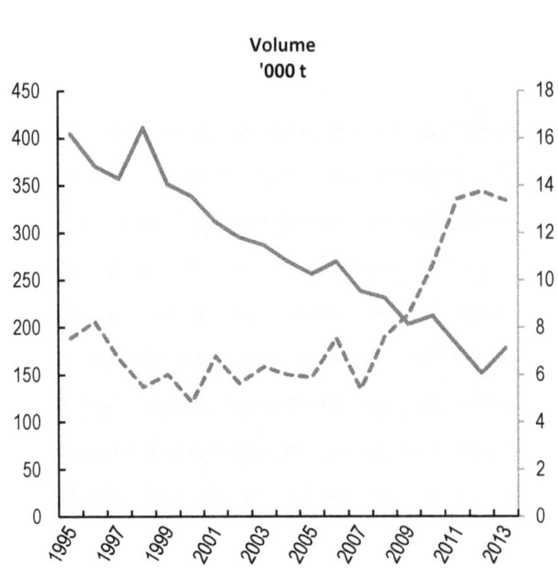

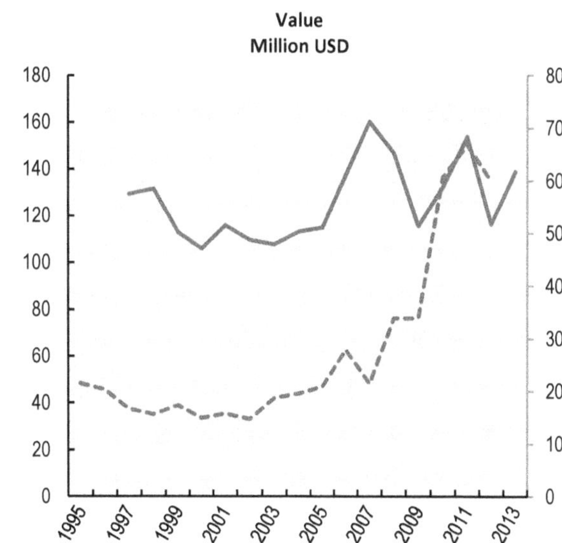

Source: FAO FishStat Database.

Key characteristics of Swedish fisheries

- In 2013, landings from marine fisheries amounted to 151 000 tonnes with a corresponding value of approximately EUR 100 million (USD 139 million). By tonnage, the most important species in the marine fisheries sprat, herring and cod. The most important species in terms of value were herring and sprat, cod, northern prawn and Norway lobster (Panel A).
- In 2013, Sweden imported a total of 688 000 tonnes of fish for a total value of EUR 3,4 billion. Exports amounted to 665 000 tonnes for value of EUR 2,7 billion. A major factor in Sweden's foreign trade is due to re-exports of Norwegian salmon (Panel B).
- Excluding fuel tax exemptions the total transfers to the fisheries sector in 2013 amounted to EUR 69 million of which EUR 64 million was general services which includes an estimation of research costs, costs for management, enforcement and control at the national level. In 2012 the value of the fuel tax exemptions was EUR 24 million (Panel C).
- 3 800 persons were employed in the fisheries and aquaculture sectors in 2012. Of these 1 389 on vessels (operating with a license), 370 in aquaculture and 1 895 occupied in fish processing. The fleet consisted of 1 378 active and inactive vessels in 2012.

Key fisheries indicators

Panel A. Key landings in value, 2013

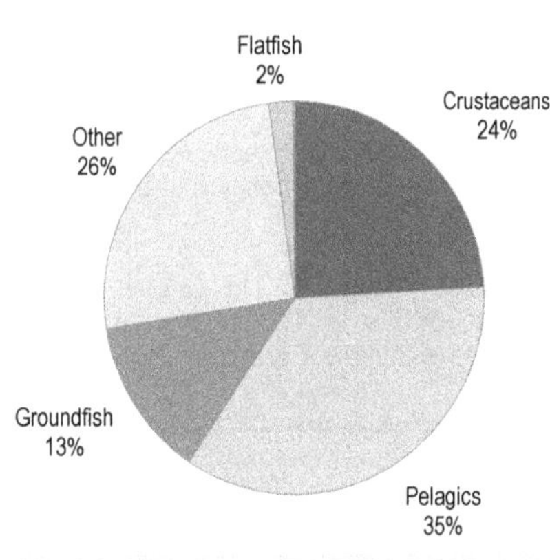

Panel B. Trade evolution

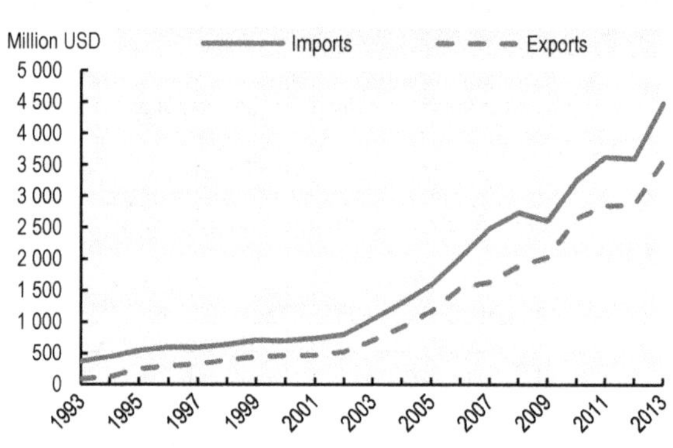

Panel C. Evolution of government financial transfers for capture marine fisheries

	Average 2005-10*	2012	% change
Number of fishers	1 796	1 572	-12
Number of fish farmers	420	370	-12
Total number of vessels	1 480	1 378	-7
Total tonnage of the fleet	40 833	30 831	-24

* Or closest available years.

Panel D. Capacity

	Average 2005-10*	2013	% change
Number of fishers	1 796	1 466	-18
Number of fish farmers	420	370	-12
Total number of vessels	1 480	1 362	-8
Total tonnage of the fleet	40 833	29 000	-29

* Or closest available years.

CHINESE TAIPEI

Summary of recent developments

- To help improve the sustainability of the fishery, the total number of large-scale tuna longliners larger than 100GRTs has been reduced from 614 to 418 by 2013. To protect sharks, Chinese Taipei promulgated the "*Regulations on the Disposal of the Fins of the Shark Catches of Fishing vessels*" and the "*Regulations on the Imports of Shark Fins*" in 2012.
- Major policy initiatives include prohibition of fisheries which deplete resources, introduction of TAC on selective stocks, installation of VMS on all deep sea fishing vessels, promotion of sustainable aquaculture and traceability of fisheries products.
- To enhance the conservation of marine and fisheries resources, Chinese Taipei issued 3 new regulations in 2013 including the Regulations on Mackerel and Scads Fisheries, the National Plan of Action to Prevent, Deter and Eliminate Illegal, Unreported and Unregulated (IUU) Fishing, and the Regulations on Glass Eel Fishing Season.
- In 2013, the annual production of Deep sea fisheries has been over 772 000 tonnes with a value of NT$43.8 billion (726 000 tonnes and NT$50.1 billion in 2012,) which accounts for over 60% of the overall fisheries production, while production of coastal and offshore fisheries is 152,000 tonnes with a value of NT$ 16.6 billion (181 000 tonnes and NT$18 billion in 2012.) Aquaculture production was 350 000 tonnes with a value over NT$ 40 billion (348 000 tonnes and NT$37.5 billion in 2012.)

Capture fisheries and aquaculture production

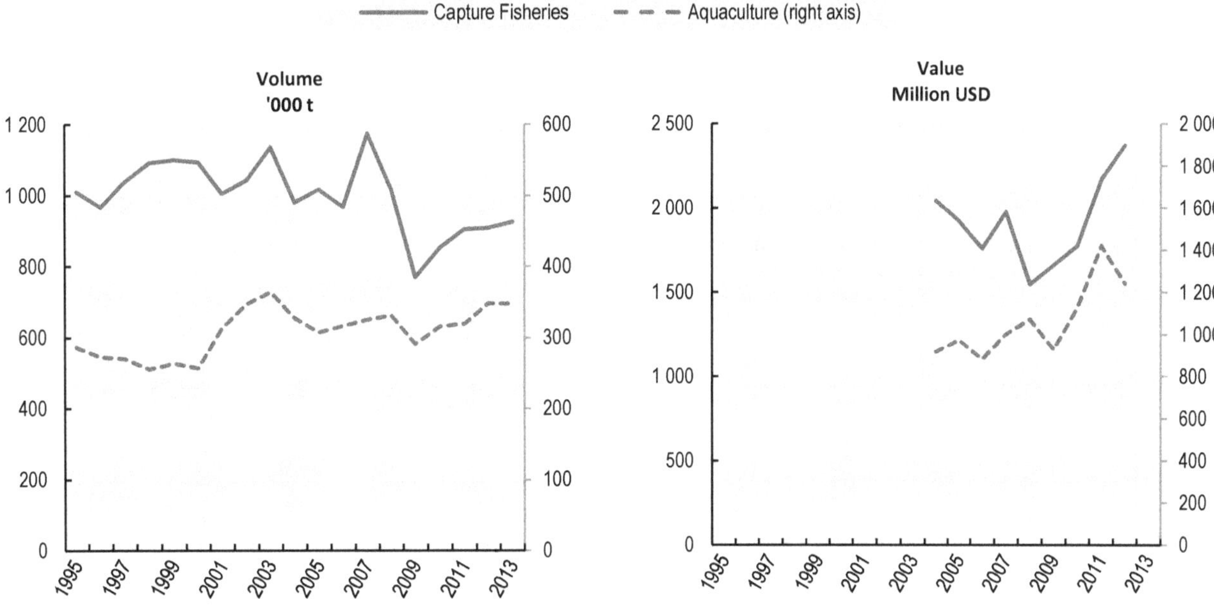

Source: FAO FishStat Database.

Key characteristics of Chinese Taipei fisheries and aquaculture

- In terms of composition of the value of landings, tuna continue to be the most important species landed in 2013, followed by pelagics, shellfish and molluscs, and crustaceans. (Panel A)
- In 2013, Chinese Taipei exported fisheries products by value of USD 1 843 million. Export value expanded by 74% during 2009 and 2012, but decreased by 10% in 2013. At the same year, value of import was USD 1 182 million. Over the last ten years, import has increased by 137% since 2003. In terms of value, the biggest export market for Chinese Taipei's fisheries products was Japan, followed by Thailand, and the people's Republic of China in 2012, while in terms of export volume, Thailand was the biggest market. China was the biggest exporter to the Chinese Taipei's market, followed by Viet Nam, Japan, Peru, and Norway. (Panel B)
- A total of USD 24.8 million was transferred to the marine capture fisheries sector in 2012, which is 44% decrease compared to the base period (USD 44.5 million). However, in 2013, Government Finance Transfer on marine capture fisheries recorded 143% increase compared to the previous year largely due to decommissioning of vessels and licenses. (Panel C)
- The number of fishing vessels and tonnage decreased 6% and 9% between base period and 2012. Compared with the average of base period and 2012, the number of people employed in the capture fisheries stayed stable with the decrease of 2% and the number of fish farmers decreased 9%. (Panel D)

Key fisheries indicators

Panel A. Key landings in value, 2013

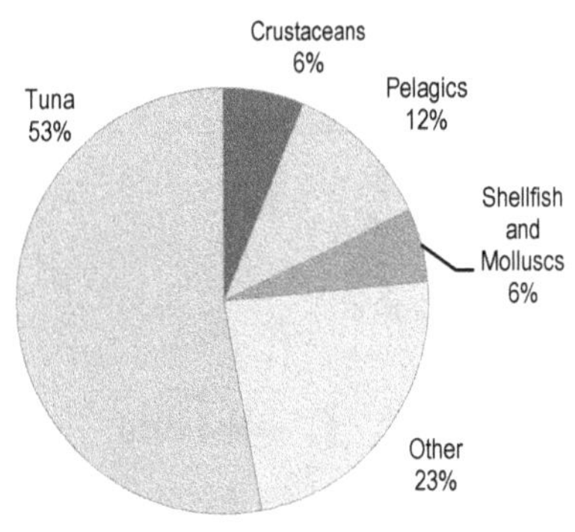

Panel B. Trade evolution

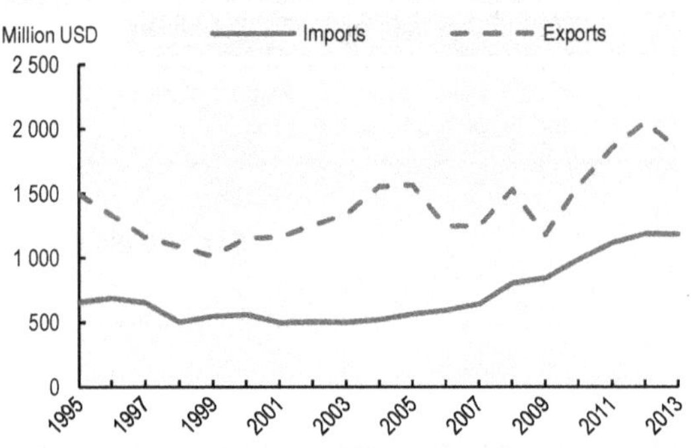

Panel C. Evolution of government financial transfers for marine capture fisheries

USD	Average 2005-10*	2012	% change
Direct payments	25 133 745	2 930 743	-88
Cost reducing	2 782 044	1 534 426	-45
General services	16 617 171	20 421 937	23

* Or closest available years.

Panel B. Capacity

	Average 2005-10*	2012	% change
Number of fishers	246 384	240 542	-2
Number of fish farmers	95 531	85 969	-10
Total number of vessels	24 927	23 446	-6
Total tonnage of the fleet	666 006	604 634	-9

* Or closest available years.

TURKEY

Summary and recent developments

- In 2013, production from marine capture, inland fishing and aquaculture were 339 047 tonnes, 35 074 tonnes and 233 394 tonnes, respectively. Compared to 2012, the volume of marine capture landings decreased in 2013 by 13.5%, which was largely driven by decrease in Atlantic bonito and Striped venus clam. The value of marine capture was slightly decreased. Total volume of aquaculture production increased from 212.410 tonnes in 2012 to 233.393 tonnes in 2013, an increase of 10%, and its value was slightly increased.
- A new Fisheries Law was prepared for the purpose of transposing EU legislations pertinent to fisheries into the national regulation in accordance with Common Fisheries Policy and submitted to the Prime Ministry for approval procedures. The establishment of the Directorate-General for Aquaculture and Fisheries in the Ministry of Food Agriculture and Livestock in June 2011 enhanced the capacity of administration as well as enforcement.
- For the sustainable fisheries, Turkish government introduced various measures including fishing vessels decommissioning scheme launched in 2012 (364 vessels in 2013), Black Sea Anchovy Stock Assessment Project launched in 2011, the National Fisheries Data Collection Programme which was put into practice in 2012, and increasing of the minimum depth limit for purse seine and pelagic trawls from 18 to 24 meters.

Capture fisheries and aquaculture production

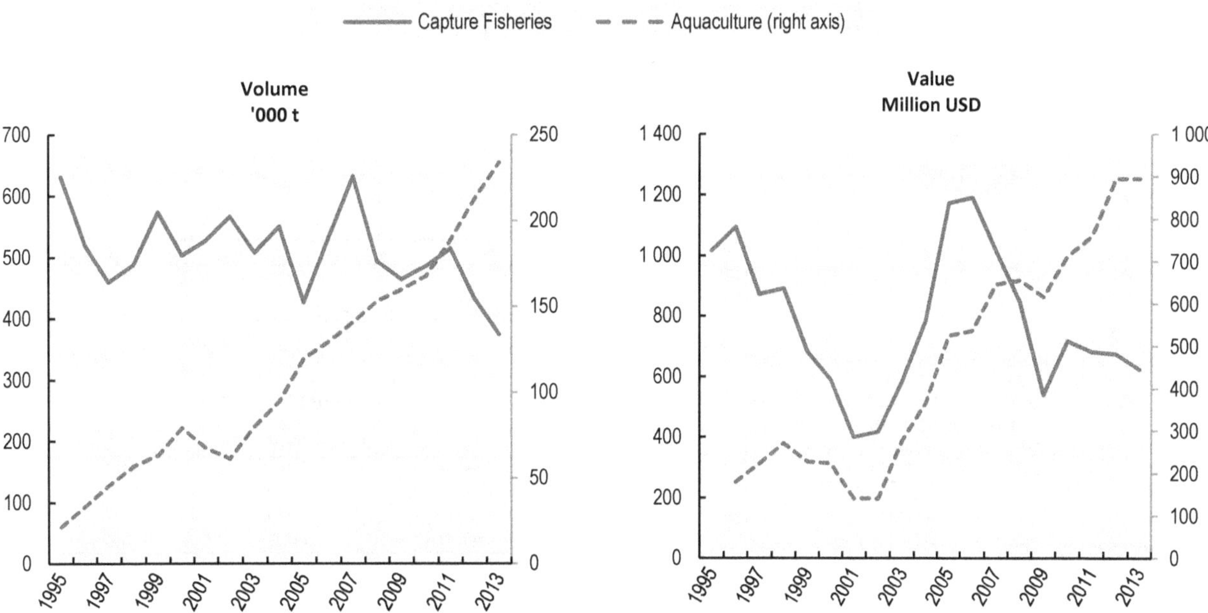

Source: FAO FishStat Database.

Key characteristics of Turkish fisheries

- Anchovy is the major species for marine capture. Anchovy accounted for 51 and 61% of total marine fish capture in 2012, and 2013, respectively. Atlantic bonito, pilchard, horse mackerel, sprat, whiting, and bluefish were other main marine commercial fish species. Two thirds of aquaculture farms are involved in rainbow trout (Panel A).
- Between 2012 and 2013, the volume of fish imports increased by 2.7% to around 67 530 tonnes. In value terms, total imports increased in 2013 to approximately 188 million USD representing 5.3% increase in 2012. Between 2011 and 2013, Norway, Morocco, United States, Georgia, People's Republic of China, and Spain were key suppliers of fishery products to Turkey. Main important species were mackerel, tuna, herrings, salmon species, cods, sardines and cuttlefish (Panel B).
- GFT for the total marine capture fisheries, aquaculture and processing sectors increased by 10% from USD 281 million in 2011 to USD 309 million in 2013. The biggest part of GFT in 2013 was general services amounting to 45% of the marine capture fisheries support. It is mainly due to enforcement service programmes and it is followed by fuel tax exemptions representing 31% of support to marine capture fisheries, i.e. 25% of total support (Panel C).
- There were 17 165 registered vessels with a total tonnage of 183 499 GT in 2011 and 16 437 vessels with 175 269 GT in 2013. The majority of fishing fleet is comprised of small vessels. In 2013, 91 percent of fishing vessels were less than 12 m in length. Thanks to the scope of the fishing vessels decommissioning scheme launched in 2012, the number of vessels of 12 m in length and over reduced 25% in 2013. (Panel D)
- In 2013, overall employment in harvesting sector (i.e. marine and inland fisheries) was 37 499 fishermen, an 8.7% percent decrease compared with 41 054 in 2012. Of the 37 499, 33 455 were employed in marine fishing sector and 4 044 were employed in inland fisheries. In 2013, there were 2 353 farms (1 935 inland and 418 marine) with 463 thousand tonnes total capacity, directly providing employment for about 9 800 people. (Panel D)

Key fisheries indicators

Panel A. Key landings in value, 2013

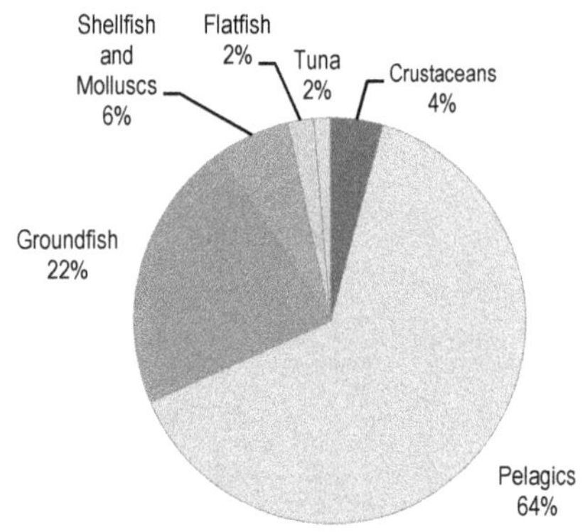

Panel B. Trade evolution

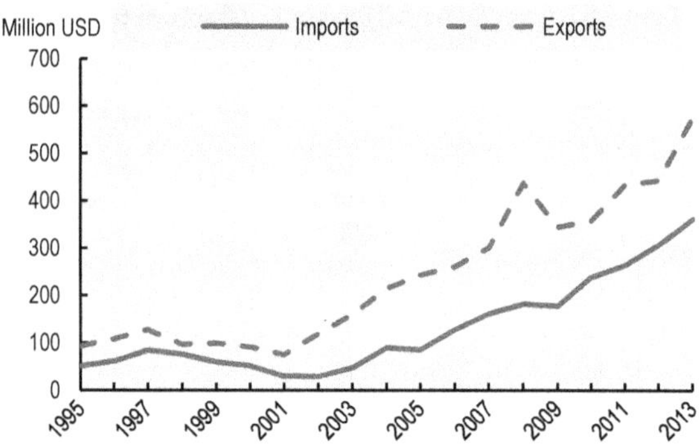

Note: Fish meal, fish oil and seaweeds are included

Panel C. Evolution of government financial transfers for marine capture fisheries

USD	Average 2005-10*	2013	% change
Direct payments	..	32 606 983	..
Cost reducing	66 835 591	77 873 965	17
General services	87 630 499	139 317 854	59

* Or closest available years.

Panel D. Capacity

	Average 2005-10*	2013	% change
Number of fishers	54 537	37 499	-31
Number of fish farmers	6 503	9 800	51
Total number of vessels	18 138	16 437	-9
Total tonnage of the fleet	188 337	175 328	-7

* Or closest available years.

UNITED KINGDOM

Summary and recent developments

- Of 14 indicator fin-fish stocks in UK waters, the proportion of stocks at full reproductive capacity and being harvested sustainably has risen from around 20% in the 1990s to around 40% in 2009-2012 (as verified by ICES). These trends are likely to be due to a combination of EU controls on total allowable catches (TACs) and of effort.
- A regulation came into force on 1 January 2014 on the Common Organisation of the Markets in fishery and aquaculture products which formed part of the package of measures under the revised Common Fisheries Policy.
- The United Kingdom focus is now on the implementation of the reformed CFP, particularly helping the UK fishing industry to prepare for the introduction of the landing obligations. Discussion with the fishing industry is on-going on the work needed to implement CFP reform, especially the introduction of discard bans, particularly for pelagic fisheries which is coming into effect in January 2015. We are working closely with the industry to make all parties aware of the changes and address problems so that the transition is as smooth as possible.
- For inshore fisheries, in particular, the question of how to extract social and environmental benefits in the context of economic liberalisation will be fundamental. The aim is for all elements of the fleet, big and small, to be economically viable and operating without long-term subsidy.
- The Marine and Coastal Access Act, enacted in 2009, will greatly improve the way the United Kingdom uses its marine resources and maximises the benefits it gets from them. Through the Act UK Administrations are now in the process of introducing new systems for marine planning and licensing within the policy framework provided by a UK Marine Policy Statement adopted by all UK administrations in March 2011. England's first marine plans (for the East inshore and offshore areas) were adopted in April 2014 and, following consultation in 2013, a National Marine Plan for Scotland is expected to be adopted in early 2015. Both Wales and Northern Ireland are working on their national marine plans and are expected to be published for consultation in late 2014 and early 2015 respectively.

Capture fisheries and aquaculture production

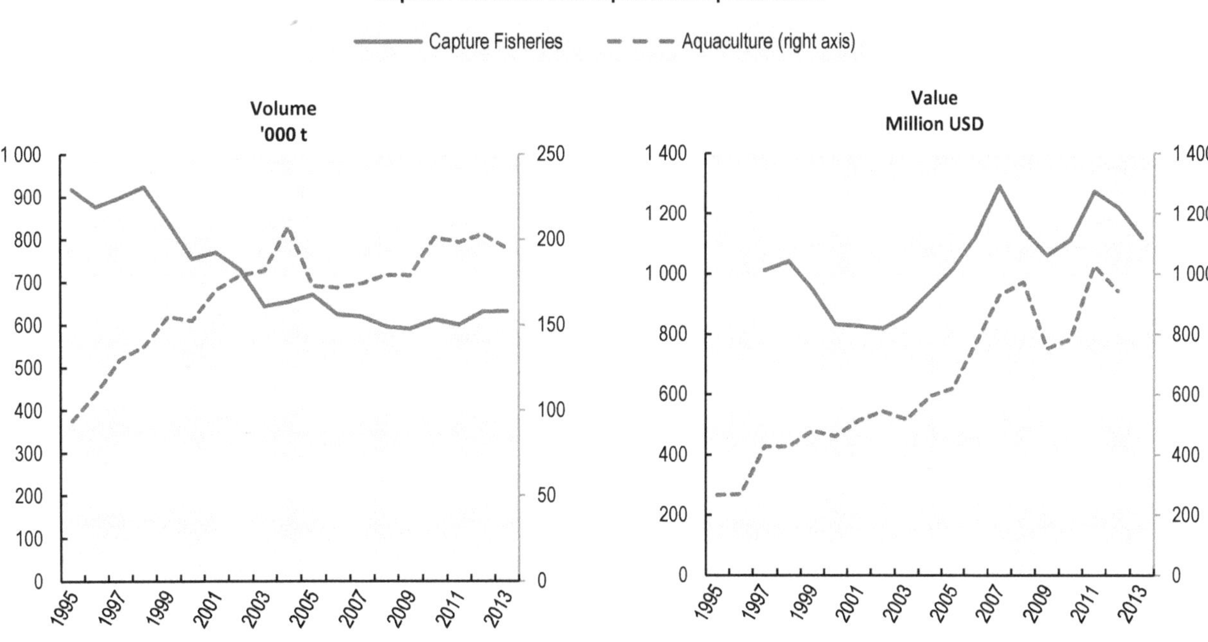

Source: FAO FishStat Database.

Key characteristics of British fisheries

- In 2013, pelagic species accounted for 25% by value of all landings by United Kingdom (UK) vessels, followed by groundfish (28%), crustaceans (23%) and shellfish and molluscs (14%) (Panel A).
- The United Kingdom remains a net importer of fish and fish products. In 2011, the value of both imports and exports have recovered from the recent off-peak to the similar level as before the economic crisis. In 2011, the value of both imports and exports increased. The value of UK seafood imports increased by 15.6% to USD 4 247 million, whilst the volume grew by 4.5% to 718 000 tonnes. The value of UK seafood exports grew by 13.7% to USD 2 395 million, while the volume fell 15% to 435 000 tonnes (Panel B).
- The European Fisheries Fund (EFF) is the main funding source for government financial transfers (GFTs). The programme runs from 2007 to 2014 and is expected to be replaced by the new European Maritime and Fisheries Fund (EMFF) in early 2015. The EMFF will run until 2020. In the 2012/13 financial year, a total of USD 18.47 million was transferred to fisheries sector from the UK Government, which is a 79.6% decrease compared with the GFTs in 2005 (Panel C).
- In 2012, there were 12 445 fishers in capture fisheries, which is a 3.4% decrease since 2007. Total number of registered vessels and total tonnage of the fleet also decreased respectively by 5.3% and 5.7% since 2007. (Panel D)

Key fisheries indicators

Panel A. Key landings in value, 2013

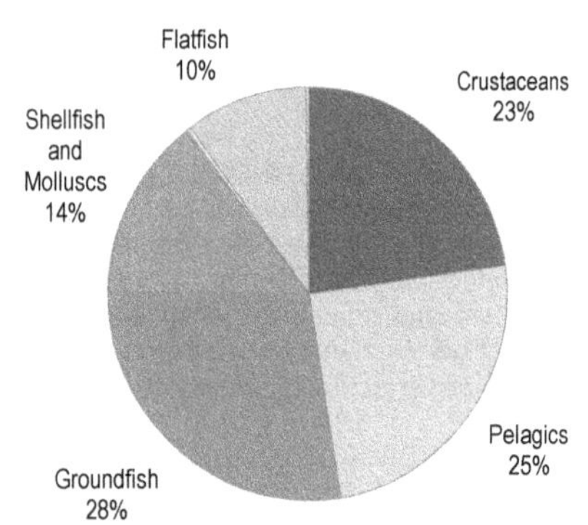

Panel B. Trade evolution

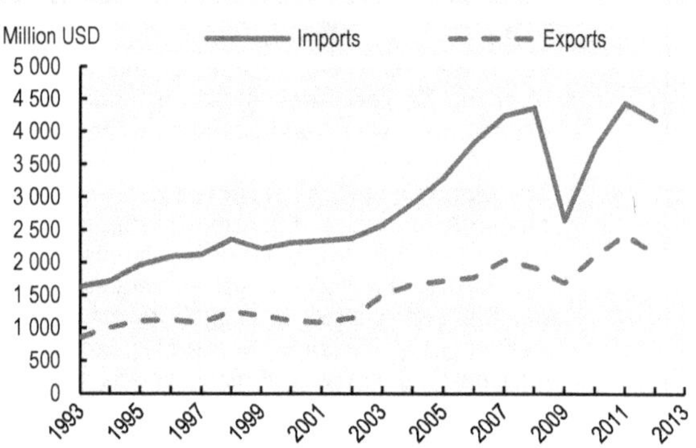

Panel C. Evolution of government financial transfers for marine capture fisheries

USD	Average 2005-10*	2013	% change
Direct payments	5 380 570	3 419 504	-36
Cost reducing	5 945 930	..	..
General services	44 798 346	11 910 321	-73

* Or closest available years.

Panel D. Capacity

	Average 2005-10*	2013	% change
Number of fishers	12 671	12 445	-2
Number of fish farmers	3 300	3 329	1
Total number of vessels	6 640	6 481	-2
Total tonnage of the fleet	210 227	198 560	-6

* Or closest available years.

UNITED STATES

Summary and recent developments

- The Magnuson-Stevens Fishery Conservation and Management Act (Magnuson-Stevens Act) provides a legal framework for addressing a wide variety of marine stewardship issues. The law mandates an end to overfishing, promotes market-based management, strengthens the role of science, improves data on recreational fisheries, and includes new measures to combat illegal, unreported and unregulated (IUU) fishing and to reduce bycatch in global fisheries.
- For 2013 the National Marine Fisheries Service (NMFS) reported on the status of 478 individual stocks and stock complexes, and determined that 9% of stocks and stock complexes with known overfishing determinations were subject to overfishing and 17% of stocks and stock complexes with known stock condition continued to be overfished.
- Two new mandates have been implemented over the last few years: annual catch limits (ACL) and accountability measures (AM). ACLs must be set at a level that overfishing does not occur in the fishery and must take uncertainty into account. AMs are corrective actions which are triggered when an ACL is exceeded. AMs may be in-season actions (reductions in effort or closures) or post-season measures taken in the following year to "payback" the overage. ACLs and AMs are now in place and effective for the 2012 fishing year in all federally managed fisheries.
- Commercial landings (edible and industrial) by United States fishers at ports in the 50 states totalled 4.4 million metric tonnes valued at USD 5.1 billion in 2012 - a 2% decrease in volume and a 4% decrease in value, respectively, compared with 2011. Alaskan pollock, menhaden, cod, flatfish, Pacific salmon and hake are the six most important species in terms of volume, while crabs, scallops, shrimp, salmon, and lobster are highest in terms of gross value.

Capture fisheries and aquaculture production

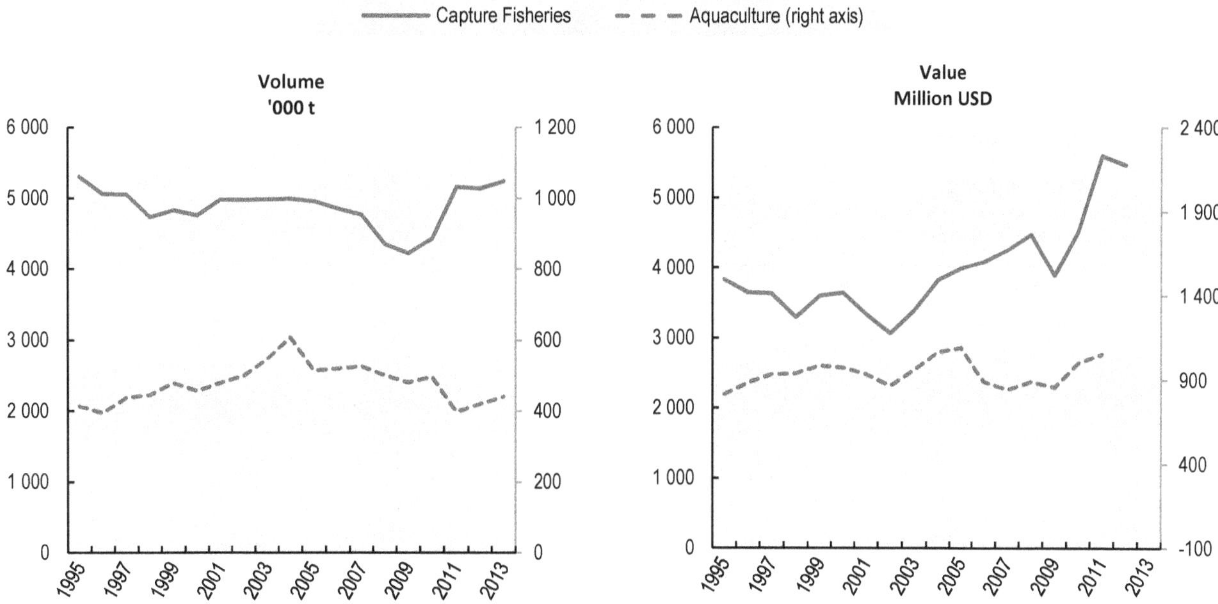

Source: FAO FishStat Database.

Key characteristics of American fisheries

- Alaskan pollock, menhaden, cod, flatfish,, Pacific salmon, and hakes are the six most important species in terms of volume, while crabs, scallops, shrimp, salmon, and lobster remained highest in terms of gross value. (Panel A)
- US imports of edible fishery products in 2012 were valued at USD 16.7 billion, essentially the same as 2011. The quantity of edible imports was 2.449 million tonnes, 1.542 million tonnes more than the quantity imported in 2011. US exports of edible fishery products were 1.497 million tonnes valued at USD 5.5 billion, essentially unchanged from 2011. (Panel B)
- In 2010, USD 1 901 million was transferred to fisheries sector from government, which is an 8.8 % decrease, compared to 2008 (USD 2 084 million). About 95% of the transfers in 2011 were spent on general services. (Panel C)

Key fisheries indicators

Panel A. Key species by value in 2012

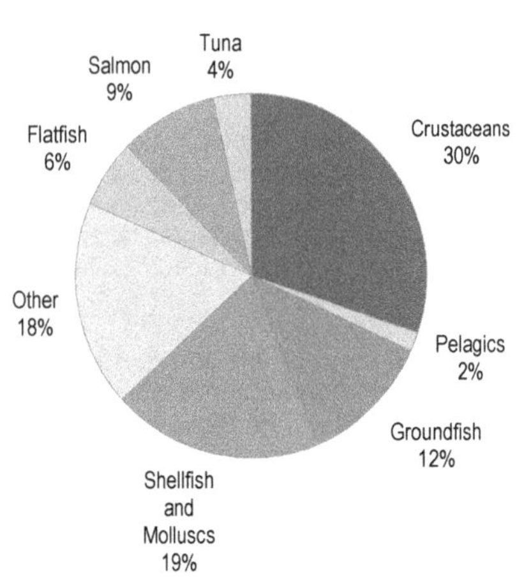

Panel B. Trade evolution

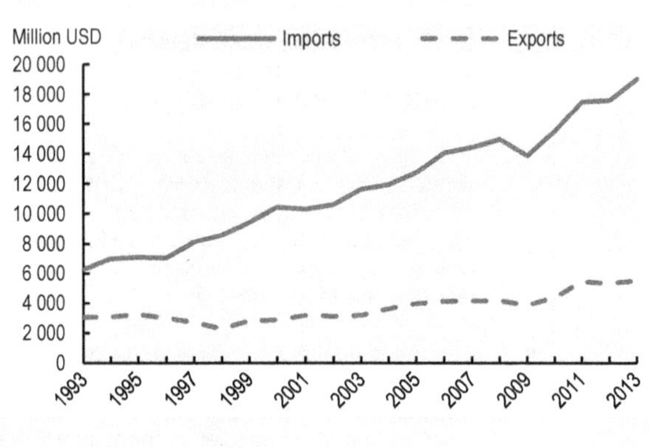

Panel C. Evolution of government financial transfers

USD	Average 2005-10*	2012	% change
Direct payments	116 221 792	..	..
Cost reducing	9 193 449	2 200 000	-76
General services	1 662 877 297	2 354 225 707	42

* Or closest available years.

ORGANISATION FOR ECONOMIC CO-OPERATION AND DEVELOPMENT

The OECD is a unique forum where governments work together to address the economic, social and environmental challenges of globalisation. The OECD is also at the forefront of efforts to understand and to help governments respond to new developments and concerns, such as corporate governance, the information economy and the challenges of an ageing population. The Organisation provides a setting where governments can compare policy experiences, seek answers to common problems, identify good practice and work to co-ordinate domestic and international policies.

The OECD member countries are: Australia, Austria, Belgium, Canada, Chile, the Czech Republic, Denmark, Estonia, Finland, France, Germany, Greece, Hungary, Iceland, Ireland, Israel, Italy, Japan, Korea, Luxembourg, Mexico, the Netherlands, New Zealand, Norway, Poland, Portugal, the Slovak Republic, Slovenia, Spain, Sweden, Switzerland, Turkey, the United Kingdom and the United States. The European Union takes part in the work of the OECD.

OECD Publishing disseminates widely the results of the Organisation's statistics gathering and research on economic, social and environmental issues, as well as the conventions, guidelines and standards agreed by its members.

OECD PUBLISHING, 2, rue André-Pascal, 75775 PARIS CEDEX 16
(53 2015 02 1 P) ISBN 978-92-64-24016-2 – 2015-17

www.ingramcontent.com/pod-product-compliance
Lightning Source LLC
LaVergne TN
LVHW081412110826
845149LV00010B/1721

9789264240162